# The Army's Nuclear Power Program

**Recent Titles in Contributions in Military Studies**

Strategic Impasse: Offense, Defense, and Deterrence Theory and Practice
*Stephen J. Cimbala*

Feeding the Bear: American Aid to the Soviet Union, 1941–1945
*Hubert P. van Tuyll*

Military Planning for the Defense of the United Kingdom, 1814–1870
*Michael Stephen Partridge*

The Hollow Army: How the U.S. Army Is Oversold and Undermanned
*William Darryl Henderson*

Reevaluating Major Naval Combatants of World War II
*James J. Sadkovich, editor*

The Culture of War: Invention and Early Development
*Richard A. Gabriel*

Prisoners, Diplomats, and the Great War: A Study in the Diplomacy of Captivity
*Richard Berry Speed III*

Military Crisis Management: U.S. Intervention in the Dominican Republic, 1965
*Herbert G. Schoonmaker*

The Persian Gulf War: Lessons for Strategy, Law, and Diplomacy
*Christopher C. Joyner, editor*

Where Eagles Land: Planning and Development of U.S. Army Airfields, 1910–1941
*Jerold E. Brown*

First Strike Stability: Deterrence after Containment
*Stephen J. Cimbala*

Legitimacy and Commitment in the Military
*Thomas C. Wyatt and Reuven Gal, editors*

# The Army's Nuclear Power Program

## THE EVOLUTION OF A SUPPORT AGENCY

Lawrence H. Suid

CONTRIBUTIONS IN MILITARY STUDIES, NUMBER 98

GREENWOOD PRESS
New York • Westport, Connecticut • London

**Copyright Acknowledgment**

All photographs contained herein are used with permission of the Office of History, HQ, U.S. Army Corps of Engineers.

**Library of Congress Cataloging-in-Publication Data**

Suid, Lawrence H.
The Army's nuclear power program : the evolution of a support agency / Lawrence H. Suid.
p. cm.—(Contributions in military studies, ISSN 0883–6884 ; no. 98)
Includes bibliographical references.
ISBN 0–313–27226–3 (lib. bdg. : alk. paper)
1. United States. Army—Electric installations—History. 2. Army reactors (PM)—History. 3. Nuclear engineering—United States—History. 4. Nuclear power plants—United States—History. 5. United States. Army Facilities Engineering Support Agency—History. I. Title. II. Series.
U6480.S85 1990
623.4'5119'0973—dc20 89–25700

British Library Cataloguing in Publication Data is available.



Library of Congress Catalog Card Number: 89–25700
ISBN: 0–313–27226–3
ISSN: 0883–6884

First published in 1990

Greenwood Press Inc., 88 Post Road West, Westport, Connecticut 06881
An imprint of Greenwood Publishing Group, Inc.

Printed in the United States of America

The paper used in this book complies with the Permanent Paper Standard issued by the National Information Standards Organization (Z39.48–1984).

10 9 8 7 6 5 4 3 2 1

To Dolores, who made this possible.

# Contents


| | | |
|---|---|---|
| | *Preface* | ix |
| | *Acknowledgments* | xiii |
| 1: | The Beginnings | 1 |
| 2: | Organization of the Nuclear Program | 19 |
| 3: | Building the Prototype | 29 |
| 4: | Fort Greely | 41 |
| 5: | The Nuclear Power Program in Full Bloom | 57 |
| 6: | The Nuclear Program in Decline | 81 |
| 7: | Transition | 95 |
| 8: | Facilities Engineering Support Agency (FESA) | 117 |
| | *Bibliographical Essay* | 127 |
| | *Index* | 129 |

# Preface

I came to the writing of the history of the Army's nuclear power program with virtually a clean slate in terms of knowledge about the subject. Like most people, I thought of the military and nuclear power only in terms of "the Bomb." Moreover, my recently published book *Guts & Glory* focused on the relationship between the film industry and the U.S. armed services, rather than on the evolution of a new military technology. However, if *Guts & Glory* remains primarily cultural and social history, the book also became military, albeit rather esoteric, history.

As an historian, of course, I viewed the contract to write the history of the Army's nuclear power program as simply a new subject to study. In this sense, the researching and writing of the book posed no unusual obstacles to overcome. In fact, to the extent that the individual armed services' public affairs offices functioned as support agencies in providing assistance to filmmakers, I had gained an understanding of the military's support operations from researching and writing *Guts & Glory*.

The Facilities Engineering Support Agency (FESA), into which the Army's nuclear power program had evolved, had accumulated virtually all the relevant documents about the origins of the nuclear power program, its operation, and its decline. FESA, of course, also had a complete documentary record of its creation and current operation. As a result, I had all the materials needed to write the history of the evolution of a military support agency. I had one additional research tool, oral history. For *Guts & Glory,* I had conducted more than 400 interviews during the period when oral history interviewing was becoming a recognized research tool. Using oral interviewing on my new project, I was able to acquire much information and understanding of the evolution of the nuclear power program. On its part, FESA was able to provide the names of people still working in the organization as well as help me to locate former members of the nuclear power program.

General K. D. Nichols, the Manhattan Project District Engineer, became the key person in providing me with an understanding of the origins of the program and the vision the Army and the military establishment had for nuclear power both in a combat setting and during peacetime. Along with General Leslie Groves and J. Robert Oppenheimer, General Nichols had the primary responsibility for building the atomic bombs that ended World War II. At the same time, even before the bombs had been successfully tested and used, General Nichols was looking forward to the peaceful use of nuclear power.

Within the collection of materials FESA gave me was an interview General Nichols had given about his role in beginning the Army's nuclear power program while head of Army Research and Development. Subsequently, I did a full-length oral history of General Nichols for the Army Corps of Engineers History Office. In the interview and later when I worked with the General on his autobiography, he discussed at considerable length his views on the use of the atom to generate electric power. As a result, I acquired an understanding of the technology as well as the problems which that technology created.

Using the full resources of FESA, I proceeded to write the history of the Army's nuclear power program and its ultimate evolution into a military support agency in the years following the Vietnam War. While I was writing under contract to FESA, no one within the agency or within the Corps of Engineers History Office ever tried to influence the organization of my work or its contents. My narrative and conclusions remain my own. If a downside to my contract existed, however, it resulted from my status as an independent contractor. I was writing the book directly for FESA rather than for the Engineers History Office. Consequently, the History Office had less commitment to the completed work than if I had been a staff historian. As a result, the Office did not give the manuscript a high publishing priority when FESA accepted it as satisfying the terms of the contract. Moreover, the agency itself had more immediate practical matters with which to deal than supervising the editing and publication of the manuscript.

Consequently, the Office did not finally agree to allow me to assume responsibility for publication of the book until the fall of 1988. Greenwood Press expressed immediate interest and the book in hand is the result. In the end, however, the delay in getting the book into print does not lessen its value and the story remains just as relevant today as it was when I completed the manuscript. The Army's desire to build and operate small nuclear power plants in remote areas showed vision for a technology that then did not exist. It also demonstrated the need for the government to take the lead in creating new technology because of costs of development by the private sector. While the implementation of the vision proved the validity of the concept, the military never came to terms with the best way to use the new technology and proponents of the program could not defend its continuation to those making the military budgets, particularly as the war in Vietnam escalated.

That the program died, however, does not negate its importance to the development of commercial nuclear power, nor did it mean that the military had solved the problems of providing electric power in remote areas. If anything, the story of the Army's nuclear power program shows that good ideas, even if they work, do not always survive.

# Acknowledgments

An author usually finds himself with a long list of people to thank for help in completing a book. In this case, however, the extraordinary time it has taken to put the history of the Army's nuclear power program into print has caused many of the names to recede from memory. Consequently, I would simply like to extend my general gratitude to all those who educated me in the whys and hows of nuclear power and so provided me with the knowledge needed to do justice to the history of the Army's nuclear power program.

At the same time, I would like to mention four people whose contributions greatly influenced the book's ultimate publication. Since the original manuscript was written long before every writer had a word processor on his desk, it became necessary to type the text into the computer to facilitate the editing process. Without the help of Dolores Haverstick in putting the manuscript onto disks and her even more important assistance of polishing the text, this book could not have appeared in its final form.

While I received many suggestions and much constructive criticism of the manuscript during and after the writing, I felt that the new version needed one more reading. As a result, I turned to two of the people who had provided me with information and then read the completed book. William B. Taylor worked within the Army's nuclear power program almost from its inception and in later years served as a consultant to the nuclear power program. He read the manuscript with considerable care from the perspective of a person who had been there and simply wanted to insure accuracy of the story.

Michael Stollmeyer was the man to whom I reported in FESA during the writing of the book and he willingly read the revised manuscript. He provided both factual and technical suggestions which insured that the story would be as accurate as possible. To both men I express my appreciation for the help and for retaining an interest in the story itself.

Finally, I would like to thank Martin Gordon for his assistance. He put me in touch with the Corps of Engineers history office soon after I returned from a year's teaching in Vermont. In large measure Martin is responsible for the book's seeing the light of day. As important as the Corps of Engineers history office archivist, he provided me with assistance during the editing process as well as with the photographs needed to illustrate the book.

# THE ARMY'S NUCLEAR POWER PROGRAM

CHAPTER 1

# THE BEGINNINGS

The birth of the nuclear age can be assigned to at least three dates. On December 2, 1942, beneath the west stands of Stagg Field at the University of Chicago, Dr. Enrico Fermi and a group of scientists of the Metallurgical Laboratory obtained the first self-sustaining nuclear chain reaction. According to Gen. Leslie Groves, head of the Manhattan Project, this marked the day "when man first released the energy of the atom and controlled it."[1]

Scientists first demonstrated the full magnitude of nuclear power when they exploded the first atomic device at Alamogordo, New Mexico, on July 16, 1945. The event moved Dr. J. Robert Oppenheimer, the father of the atomic bomb, to recall a line from the sacred Hindu epic *Bhagavad Gita*: "I am become death, the destroyer of worlds." The whole world became aware of the destructiveness of nuclear energy on August 6 and August 9, 1945, when the U.S. Army Air Corps dropped atomic bombs on Hiroshima and Nagasaki and so ended the war. To George Caron, tail gunner on the *Enola Gay*—which carried the " Little Boy" on the first atomic mission—the bomb's shock wave looked as if "the ring around some distant planet had detached itself and was coming up toward us." Capt. Robert Lewis, the plane's co-pilot, exclaimed, "My God, look at that sonofabitch go!" and not "My God, what have we done?" as historians have often credited him as saying. Nevertheless, his famous comment, actually written in the log of the B–29, "probably reflected the thoughts of all 12 men aboard the *Enola Gay*," according to Gen. Paul Tibbets, the plane's pilot and commander of bomb delivery group.[2]

Even before the first bomb dropped on Japan, however, President Harry S Truman had directed his thoughts toward nonmilitary uses of the atom. Following the successful test detonation and the start of the *Enola Gay's* mission, Truman wrote in his diary on July 25, 1945, "It is certainly a good thing for the world that Hitler's crowd or Stalin's did not discover this atomic bomb. It seems to be the most terrible thing ever discovered, but it can be made the most useful." In announcing the bombing of Hiroshima on August 6, the President said the atomic bomb had "added a new and revolutionary increase in destruction to supplement the growing power" of the nation's armed forces. He warned that the United States was now "prepared to obliterate more rapidly and completely every productive enterprise the Japanese have above ground in any city.... Let there be no mistake, we shall completely destroy Japan's power to make war."[3]

In closing his statement, however, the President turned to the positive aspects of the new weapon. He explained that the atomic bomb also "ushers in a new era in man's understanding of nature's forces." Although a long period of "intense research" was needed, Truman prophesized that in the future, atomic energy would supplement power that came from coal, oil, and falling water. To control the production and use of atomic energy within the United States, the President said he would ask Congress to establish an appropriate commission and would later make recommendations on how "atomic power can become a powerful and forceful influence towards the maintenance of world peace."[4]

Following the bombing of Nagasaki, which forced Japan to accept the inevitable end of hostilities, Truman acknowledged "the tragic significance of the atomic bomb." Recognizing the threat that the bomb posed to "a lawless world," Truman proclaimed that the U.S. citizens "must constitute ourselves trustees of this new force—to prevent its misuse, and to turn it into the channels of service to mankind." In fact, even during the war, the scientists at the Metallurgical Laboratory, located first in Chicago and then at the Argonne Laboratory, were allowed and encouraged to start exploratory work on nonmilitary uses of atomic power.[5]

With the war over, many officers in the armed services believed that economic considerations dictated that the military would utilize nuclear power to generate electricity before any commercial application. In particular, the first two postwar Chiefs of Engineers, Lt. Gens. Raymond A. Wheeler and Lewis A. Pick, had supported the Armed Forces Special Weapons Project (AFSWP), the nuclear weapons program, and other atomic energy operations on this premise. According to Maj. Gen. Kenneth D. Nichols, who had served as Manhattan District Engineer and later had succeeded his former boss, Lt. Gen. Leslie Groves, as Chief of AFSWP, both Wheeler and Pick believed that "ultimately atomic power would be a reality and the Corps of Engineers would have a specific requirement for such atomic power and for a pool of officers with a background of experience in atomic energy."[6]

Despite such interest in developing nuclear power within the armed forces and the probable feasibility of the technology to generate electricity from the atom, the first postwar military application came from the combat side of research and development; the Navy began to develop a nuclear submarine program in 1948. Only after the Army created its Office of Research and Development in the fall of 1951, and appointed Gen. Nichols as the Office's first Chief in January 1952, did the service begin to develop nuclear energy for nondestructive military purposes.[7]

As Manhattan District Engineer and later as Chief of the AFSWP until late 1950, Gen. Nichols had long been committed to developing atomic energy to generate power. His new assignment as Chief of the Army's Research and Development Office provided him with the platform from which to implement his idea. In addition, Nichols served as the Army representative to the Atomic Energy Commission's Military Liaison Committee (MLC). The MLC, created within the original Atomic Energy Commission (AEC) bill, functioned generally on the level of the Joint Chiefs of Staff. With members from each of the armed services, the MLC advised the AEC what the Defense Department wanted in the way of atomic energy activities. As a result of his long involvement in nuclear matters and his current assignments, Nichols had ready access to the Secretary of the Army, the Army Chief of Staff, and other officials who could assist him in fulfilling his commitment.[8]

Once settled into his new position, Nichols quickly took steps to explore the prospects for the Army of utilizing nuclear energy to produce heat and electricity. Following some preliminary study meant to provide him with data on potential areas of investigation, on May 21, 1952, Nichols visited Dr. John Dunning, Dean of Engineering at Columbia University. During the meeting, Nichols discussed the possible application of atomic power to Army requirements. He explained to Dunning and other faculty members that Army support of the development of a small power generator might be of considerable value in assisting the ultimate development of atomic power for commercial use. Nichols did acknowledge that the Army's requirement would not have a top priority. Nevertheless, he suggested that the combination of development for potential military use with the prospect that this development would give impetus to further development for commercial use would make the project attractive to a university research group.[9]

In response, Dunning indicated the subject should be approached through a study contract that would explore potential Army uses of nuclear power, consider the state of the art as it is related to the military's needs, and make overall cost analyses to determine whether any type or size of atomic reactor would be competitive with other possible sources of power for specific Army uses. Given the number of high-priority research projects that Columbia University already had, Dunning explained that his department would not be interested in undertaking the study unless the contract also gave consideration to the long-range commercial utilization of atomic power. His staff did advise Nichols that it could foresee a nuclear energy capability that would give the Army a source of power it could utilize in isolated regions while also making a major contribution to the advancement of reactor development.[10]

With this assessment in hand, Nichols began discussions within the Army and the Corps of Engineers to begin the research and development process. Secretary of the Army Frank Pace, Jr., took the first concrete action to implement Nichols' efforts on July 9, 1952, when he wrote a Memorandum to the Army Chief of Staff Gen. J. Lawton Collins about the military application of atomic power. Referring to previous discussions and memoranda on the need to broaden the Army's initiative in the atomic energy field, Pace explained that he had "long felt that we have a duty and a responsibility to explore vigorously all avenues of promising atomic applications which could result in increasing fire power, mobility and operational flexibility and in decreasing our logistic burdens." Noting that the Army had so far devoted its nuclear efforts to warhead applications, he suggested that "the time is now upon us when it is incumbent that we concurrently apply ourselves to other applications."[11]

With the Air Force and Navy already engaged in the use of atomic power for propulsion in the air and on the sea, Pace outlined the fields in which the Army "would have a compelling and legitimate interest." The Secretary gave primary importance to the development of static nuclear power plants that "would find specific application in a logistic back-up of land forces, particularly where conventional power is not applicable." Pace thought such atomic generators could ultimately outstrip "more cumbersome and time consuming installations of conventional power generating equipment." If these generators could be designed with mobile platforms, he concluded that "the greatest gain would perhaps lie in the reduction of bulk fuel requirements."[12]

Pace then provided a survey of other potential areas of Army involvement in nuclear research and development, arguing that the service "should proceed with vigor and incisiveness and that to do so it will be necessary, not only to adopt appropriate personnel policies of an affirmative nature, but also to create a suitable organizational element effectively to deal with these opportunities." The Secretary did not think the Army could rely on personnel inherited from the Manhattan Project for any future programs. Instead, he recommended that even before the Army formulated a nuclear program, "as a minimum we should associate a number of officers with the AEC Reactor Division and send certain officers to the AEC Reactor Technology School pending the time when we attain some momentum." In closing, Pace asked Collins for "your views and your suggestions at an early moment as I know you agree with me as to the military importance of this whole subject."[13]

On his part, Gen. Nichols continued his inquiries into the organization of an Army nuclear power program. On July 10, 1952, he had a meeting with Brig. Gen. Gordon Textor of the Chief of Engineers' Office to discuss involving the Engineers in the program and assigning an engineer officer to become project officer for the duration of the project. Textor indicated that the Corps would cooperate and said he would discuss the project with the Chief of Engineers Lt. Gen. Pick. The same day, Nichols called Dr. Lawrence Hafstad, Head of the AEC Reactor Division, to advise him of the meeting and the conclusion to give the Engineers the job of directing a study group such as the one proposed by Dr. Dunning. Hafstad agreed with the approach, indicating he would prefer that the AEC not handle any project studies. He also said that if the Engineers would assume the research effort, he would put a mutally agreed upon Engineer Officer in the Reactor Division in order to keep him abreast of current AEC developments in the reactor field.[14]

Following word the next day from Textor that Gen. Pick "was in perfect agreement all the way through on the atomic power project," Nichols drafted a Memorandum to Gen. Collins on July 15, outlining the background of the Army's involvement in developing atomic energy as a source of power. He explained that until recently "the state of the art did not indicate the desirability for the Army Corps of Engineers to sponsor aggressively a land based atomic power project." He then described his efforts to reassess the Army's position since he had become the Chief of the Army's Research and Development Office in January.[15]

In the Memo he sent on July 22, 1952, Nichols summarized his discussions with Dunning and Hafstad and reported on his conversations with industry representatives at recent meetings of the Nuclear Energy Application Committee of the American Society of Mechanical Engineers. He said that the power industry believed "the time is approaching when they should enter vigorously into the development of a power plant using an atomic reactor as a source of heat and as a source of replacement fuel," i.e., a breeder. This belief on the part of the conservative representatives of the power industry is a new development in the field of atomic power.[16]

Nichols cited two deterrents to private industry's proceeding "aggressively" to develop a nuclear power plant. The Atomic Energy Act of 1946 created an obstacle to industry's full participation in an effort financed by private capital. At the same time, the lack of nuclear engineering expertise made private investment "a very considerable risk." According to Nichols, as an alternative to the problems involved in a private approach, industry representatives proposed that the government build a nuclear reactor with a primary purpose of producing fissionable material for military use and secondarily to provide power for private sale. Nichols explained that this suggestion would make the power industry dependent on government support and subsidy. In contrast, he said that although the private approach would require more time, it represented the path that many people believed was required to create "a large, self-sustaining atomic power industry in the United States." According to one knowledgeable opinion Nichols had received, "this more desirable approach" would require about three years "to advance the state of the art to such a degree that the engineering risk will be more compatible with private capital risk." Nichols also noted that it would take the same time "to achieve changes in the Atomic Energy Act that would permit more vigorous participation by industry."[17]

In light of his findings, Nichols said he had explored with qualified people the potential of an Army program "based on satisfying a potential Army requirement for a mobile or semimobile atomic power plant." He thought it obvious that such a program "would make a major contribution to advancement of the state of the art and to an ultimate industrial power industry, because the Army sponsored effort would permit proceeding vigorously prior to amending any restrictions imposed by the Atomic Energy Act, and, further, the initial engineering and economic risk inherent in such an enterprise could be absorbed justifiably by the Army." At the same time, Nichols expressed his "personal conviction" that the Army had various needs "for a mobile or semi-mobile power plant that is essentially free of the present logistic support requirements for fuel." Consequently, he concluded that "now is the time to put the Army actively in the atomic power program."[18]

Nichols saw his exploratory inquiries and initial research since becoming Chief of Research and Development as the "preliminary steps" to set up a long-term program designed to satisfy such a requirement. He outlined this proposed program in a draft Memorandum he had prepared for Collins' reply to Secretary Pace's Memorandum of July 9. In requesting the Chief of Staff's approval, he advised Collins that Gen. Pick remained "in agreement with this approach."[19]

In the approved Memorandum, dated July 22, 1952, Collins assured Secretary Pace that he "strongly" concurred that the Army should have a "dynamic program" in the area of nuclear energy. To accomplish this, the Chief of Staff proposed that the Army immediately enter into a contract with a qualified research group to begin work on a program that would produce a mobile or semimobile atomic power source within a four- to eight-year period. The Memo acknowledged that the cost of producing power from the first nuclear plant would probably exceed that from a conventional power source. Consequently, the Army would have to justify its expenditure of funds on the basis of logistic savings in transporting fuel to remote areas. Secondarily, the service would have to justify the program as providing an impetus for the development of the commercial nuclear power industry and so offer great potential value to the nation as well as having "a major indirect military benefit."[20]

The Memorandum went on to explain that the research contract would not only be feasible, "but very timely in light of the progress made to date in reactor development and in light of the studies that are being made by the power industry." To initially finance the Army's Nuclear Power program, Collins proposed that the Army provide approximately $250,000 to the Chief of Engineers to administer the contract. During the 1953 fiscal year, Collins said that the Chief of Research and Development would need an additional $1 million to continue the Program, to be obtained either by reallocation of his budget or by appealing to the Secretary of Defense's emergency fund. According to the Memo, Dr. Hafstad agreed that the Corps of Engineers should initiate the project. Once it had developed a firm military requirement and defined a specific reactor project, the Army would consult with the AEC "to determine the best way for proceeding with the engineering and construction phase."[21]

Collins explained that the initial contract would be expected to accomplish three goals. It would determine the nature of the Army's requirements for power sources to be provided by power plants. It would ascertain the most effective type of reactors to meet those requirements and provide a preliminary design analysis to show the scope of the effort required. Finally, it would arrive at a proposed program that would include a probable time schedule, estimated cost, and organizational structure recommended to carry out the project.

The Chief of Staff singled out the immediate need to assign an officer to the project who could carry it forward with "dedicated vigor, determination and continuity." He reported that the Chief of Engineers had agreed to assign an officer to the program for as long as needed. Collins said he was prepared to approve and he urged Pace to recognize that "the officer selected for this position shall be permitted to remain with it so long as assignment appears desirable." On the other hand, decisions regarding size of the support staff and whether it should include liaisons at various AEC reactor sites could be deferred until the completion of the initial contract.[22]

In concluding, Collins raised the issue of how the Program would be classified for security purposes. He "earnestly" proposed that the project be unclassified except for details "that must be classified in accordance with the regulations of the AEC as required by the Atomic Energy Act." Collins justified this position by voicing a theme that would run through the life of the Army's Nuclear Power Program. It offers "a fruitful source of effective publicity on Army support of peaceful application of atomic energy."[23]

While Collins' Memo was passing upward through the chain of command, Gen. Nichols continued to establish the groundwork for the future Nuclear Power Program. On July 17, Dr. Dunning called to say that Columbia "would be very interested in doing the job but could undertake it only if it is to be a long-range thing." He explained that the school would have to reassign people to the contract from other work, from other institutions, and from other contracts with industry. Since Columbia had a firm policy that it would not handle any project on a subcontract basis, Dunning suggested that if the Army wanted the university to work with a private research firm, it would have to give prime contracts to both and have it cross-referenced to each.[24]

On July 29, 1952, Secretary Pace responded to Collins' Memo by giving the Chief of Staff's proposal his "complete approval and support." He expressed his gratification at "the specific steps proposed, or already taken, to accelerate the Army's participation and action in this most promising area of research and development." He hoped that Collins could select a contractor "at an early date" to initiate the research phase of the Program. The Secretary also gave "hearty endorsement" to the concept of selecting an officer to run the Program on a long-term basis, calling it a "most forward-looking one."[25]

In conclusion, Pace wrote that he agreed the Army "should strive to accomplish this program on an unclassified basis as you propose. The far-reaching benefits of this program in terms of peaceful application should be made known as fully and as rapidly as possible to the American public." He also asked to be kept "closely informed as to the progress in the vigorous execution of this program." To this end, he suggested a meeting with Collins on the subject in early October. In the meantime, he said his office would be available to provide assistance "in the furtherance" of the project.[26]

On August 4, responding to the Secretary's approval, the Chief of Engineers Gen. Pick, with Gen. Nichols' concurrence, selected Col. James B. Lampert to head the Army's Nuclear Program. Lampert had first become involved with atomic power in 1946, when, as a promising young engineer officer, he received an assignment to the Manhattan District as a Corps of Engineers' replacement to the wartime staff. A long tour of duty had enabled him to become familiar with atomic energy activities. Subsequent assignments as District Engineer in Charleston from 1949 to 1950, and in a civil works capacity in Tulsa at the time of his new appointment, had given him administrative experience. Arriving in Washington in the middle of August for his initial briefing, Lampert found the Program "was on an extremely tentative state."[27]

In their first meeting, Gen. Nichols told Lampert he hoped the Army's Nuclear Program would follow the pattern established by Rear Adm. Hyman Rickover in developing the Navy's nuclear submarine program. Rickover held a position as an Assistant Chief of the Navy's Bureau of Ships and as a head of an operating program within the Atomic Energy Commission. In addition, the then-captain operated with a Navy budget and an AEC budget to finance the nuclear submarine program. Although Lampert ultimately created a similar organization, he first had to overcome resistance within the AEC to "the idea of establishing another military program within [its] bosom."[28]

In the meantime, as Lampert was beginning his initial exploration of his new assignment, the Army formally designated the Corps of Engineers as the supervisory agency. By the time the Colonel had completed his move to Washington in early September, the Corps of Engineers had begun discussions with the Atomic Energy Commission on procedures for getting the program quickly underway. The AEC "recommended very strongly" that the Army follow the same procedures that the Navy and Air Force had used with their nuclear reactor programs. Like his counterparts, Col. Lampert would receive dual assignments, as head of the Army's program and a position in the Reactor Division of the AEC. This arrangement would enable him to be in immediate touch with all aspects of the Commission's reactor development program. The AEC also suggested that the Army officially express to the Agency its interest in starting a reactor program so that the AEC would have a basis for active cooperation in the Army's program.[29]

To accomplish the last item, Secretary Pace wrote to Gordon Dean, Chairman of the Commission, on September 10, 1952, outlining the Army's interest in developing a nuclear power capacity. According to Pace, at some point in the future the Army would have sufficient need for a power source free from conventional fuel requirements that would justify the additional cost of a nuclear power facility over a conventional power plant. The Secretary asked the Atomic Energy Commission for "advice and assistance" in selecting a contractor for the feasibility study and cooperation with the research group so that the study would benefit from the work of the AEC's Reactor Division. Pace also asked that the Commission accept Col. Lampert on a dual assignment basis in its Reactor Division.[30]

In his response on October 7, 1952, Dean agreed to cooperate with the Army in its study of possible military uses of nuclear energy. He pointed out that the AEC had already done studies on the economics of operating small nuclear reactors. As a result, he thought the Army should avail itself of the information on hand before extending a contract for an independent feasibility study. In fact, the AEC Chairman thought the Commission should retain responsibility for the study of any new reactor design to avoid duplication and confusion. Dean did agree that Lampert should work closely with the AEC's Division of Reactor Development in evolving any Army reactor program.[31]

Meanwhile, Gen. Nichols had been working with Col. Lampert to establish a viable foundation for the Army's program. On September 17, Dr. Dunning had come down from Columbia University to go over the project with them, and on September 22, they had met with Thomas Murray, one of the AEC Commissioners, to discuss the proposed program. Murray followed up the meeting on October 9, when he called Gen. Nichols to find out if Dean's letter to Secretary Pace "fit the bill." He also suggested Nichols and Lampert meet with him again to talk about the mechanism of the Army's program after Lampert had spent a week in the Commission's headquarters. In turn, on October 16, Nichols advised Pace that Dean's letter "may furnish an adequate basis for initiation of the program" even though it contained "certain reservations" the AEC had in "relinquishing any jurisdiction in the reactor field."[32]

Strategically, Nichols did not think the Army "should raise any issues concerning procedure and justification at this time." Instead, he thought Lampert could work directly with Dr. Hafstad to develop plans for the best means of initiating the Program. In fact, Nichols informed Pace that Lampert had already established a liaison with the AEC through Hafstad and had begun his work. He also provided the Secretary with a draft reply to Dean's letter, which advised the AEC Chairman of the arrangement.[33]

Initially assigned in the Chief of Engineer's Office to the Assistant Chief of Engineers for Troop Operations, Lampert had begun working with Hafstad in the AEC's Reactor Development Division on an informal basis. Hafstad provided him with a desk in the Civilian Reactors Branch in order to give him access to nuclear experts both in the government and in private industry. Lampert also quickly approached Adm. Rickover but obtained an interview only with some "difficulty." As Lampert recalled, the Admiral greeted him by saying, "I understand you want to build nuclear power plants for the Army. My advice to you is that you don't know what you're getting into and the best thing you can do is get out of it in a hurry." Despite the coldness of his introduction, Rickover did provide Lampert with a great deal of technical materials from the Naval Reactors Group Library and had one of his assistants brief him on the progress of the Navy's reactor program.[34]

For the most part, Lampert's first months of work consisted of a series of data-gathering meetings. At the AEC, he continued to encounter resistance in part because the Army still had not demonstrated a formal requirement for a nuclear power plant and the Commission was very reluctant to allocate funds without such a justification. At the same time, because the AEC was encouraging private industry to begin work on commercial nuclear power plants, Lampert was told to wait until he could use the results of their investigations. From other people whom he met, however, Lampert found a "rather affirmative attitude." The consensus of opinion was that the Navy's reactor program offered no immediate economic payoff in the field of energy production. Consequently, there was an interest in seeing another government nuclear program started that would offer the potential for the development of military nuclear power plants on land and could contribute to the advance of commercial nuclear power.[35]

Apart from gathering information and receiving encouragement to move ahead, Col. Lampert's efforts received a boost when he obtained an assistant, Cap. William B. Taylor, who had been working in the Armed Forces Special Weapons Project. Arriving in October, Taylor initially devoted his attention to studying the military procedures for starting a new technical program. He and Lampert also began surveying the Army's power requirements as well as the power requirements of the other armed services for which the Chief of Engineers had responsibility. By November 10, when the Nuclear Power Division was formally activated within the Office of the Chief of

Engineers, Lampert and Taylor had concluded that the key to getting the Army's Program underway was to get an official statement from a high level in the Pentagon stating that a military requirement existed for the development of nuclear power plants. According to Lampert, this "charter," based on the early Army memoranda and his many discussions, proved "invaluable... because I was never bothered by anybody in somebody's outer office trying to say, 'You can't talk to my boss because you haven't come through channels.' I think that if I had not had that sort of a general license to operate freely, it would have been much more difficult than it was to get started."[36]

While Lampert and Taylor were conducting their preliminary technical and feasibility studies, Gen. Nichols continued his own efforts to advance the Army's Nuclear Program. Following the 1952 presidential election, Nichols visited Adm. Louis Strauss in New York on December 3, to describe his initiatives toward developing an Army atomic power project and to seek Strauss' support after President Eisenhower's inauguration. Strauss, one of the original members of the AEC and then an adviser to the President-elect on nuclear affairs, gave his "hearty concurrence"" to the Army's plans. Later, this interest and encouragement translated into valuable assistance when the Army's Program ran into serious AEC objections in the spring of 1953.[37]

In the meantime, Lampert and Taylor pursued their research with the assistance of the Engineer Strategic Planning Group, the Operations Research Office of John Hopkins University, and several engineer officers, who had relevant scientific backgrounds and were on temporary assignment in the Chief of Engineer's Office. The survey of military requirements focused on power needs in remote areas—in the Mediterranean and the Middle East—the feasibility of floating nuclear power plants, and the use of mobile power plants for direct support of tactical units. To Lampert and Taylor, the most convincing argument in favor of military nuclear power plants related to the logistics of producing electricity and steam in remote areas. As Lampert saw it, once a power plant was in place and running, the military "would then have electricity practically indefinitely and [it] would not have to bring any fuel to keep the plant going."[38]

To justify the requirement to the AEC, the two officers decided to pick out locales where the Army was then generating electric power, but at high cost and with great difficulty due to the transportation problems. Their preliminary comparisons between the cost of conventional power and nuclear power led them to look specifically at military bases in such places as Northern Greenland and Alaska, where the logistical problems of delivering diesel fuel suggested that nuclear power plants might be a viable alternative. Because the Air Force had bases in such remote areas, it became involved in the study and supplied the Nuclear Power Division with much of its preliminary logistical information. Although the Navy ultimately expressed an interest in the Army's Program because of its base in Antarctica, the service provided few leads during the initial research phase because it's overseas bases were accessible to the sea and usually in moderate climates.[39]

With Lampert's preliminary studies well under way, on February 2, 1953, the Army Nuclear Power Division issued an Interim Report titled *Army Nuclear Power*. In it, Lambert and Taylor concluded that a military requirement did exist for electric power plants for use at remote bases and in prospective theaters of operation. Their research indicated that nuclear power plants might be technically feasible as well as both militarily and economically justified to fill a specific portion of this requirement. Although they acknowledged that the construction cost of a nuclear power plant would be more than that for a conventional plant, diesel or oil delivery costs could exceed the first-cost differential after only a few years of operation at some remote military bases. They said that further study was needed to ascertain the required characteristics of military nuclear power plants and this could be accomplished through the current research. In addition, they recommended that a technical feasibility study of reactor concepts and designs be undertaken in order to select one concept for development and to set forth the building program. They suggested that this study should be done, preferable on a contract basis in the "immediate future."[40]

Lt. Gen. Samuel D. Sturgis, the new Chief of Engineers, approved the Report and sent it up the chain of command to the Military Liaison Committee. The Committee concurred with the findings in the Interim Report and transmitted it to the AEC on March 23, 1953. Moreover, the new Secretary of the Army, Robert Stevens, advised the Chief of Staff on April 13, that he "wholeheartedly" subscribed to the Corps of Engineers' objective to "explore vigorously all possibilities of atomic application in meeting military requirements." Consequently, he offered "full support to all programs" in the field.[41]

In turn, Gen. Nichols helped prepare the letters that formally established Col. Lampert's position and authority and then wrote the letter to Gordon Dean and the AEC, which accompanied the Interim Report. Nichols recalled that he wrote the letter "in such a fashion that I felt that it would not be turned down." It proposed that the Army offered to pay for the nuclear project, let the AEC pay for it, let the AEC run it, let the Army run it, or run it in the way Adm. Rickover was running the Navy nuclear submarine program. Nichols recalled that the Army "did not prescribe any procedure for doing it, with the idea that we did not want to get into an argument with the AEC about how it be done or how it be funded." Nevertheless, the Commission rejected the Army's proposal, according to Nichols, because some of the Commissioners were worried that the Engineers were getting back into atomic power.[42]

In analyzing the situation in order to respond to Secretary Stevens, Brig.

Gen. Kenner F. Hertford, Assistant Deputy Chief of Staff for Research and Development, concluded that the turndown "apparently arises from the current importance of the general industrial atomic power picture in both the executive branch and the Congress as well as from a feeling that the AEC must keep the initiative in this field." At the same time, Hertford cited President Eisenhower's statement in a National Security Council meeting in late March in which he said that he didn't want Defense Department funds used to develop nuclear power plants that could be used for industrial purposes and so compete with present commercial power plants. Hertford also mentioned the controversy over the future course of nuclear power development and suggested that until the Legislative and Executive branches of the Government had established a national policy, it was "considered inadvisable" to ask the AEC to take any further action in regard to the Army's program.[43]

Despite the reluctance of the AEC to allow the Army to become involved in nuclear power, the Commission did direct Dr. Hafstad to push studies of small reactor designs that would be applicable to the Army's interest. As a result, Hafstad was proceeding with a technical study at the Oak Ridge National Laboratory, which would provide analyses to meet the Army requirements and to which Lampert would have ready access because of Hafstad's "strong personal support" of the Engineers' investigations. Consequently, although the Army's proposed research contract to study the feasibility of a military requirement for nuclear reactors could not be given out, Hafstad's Oak Ridge Study was expected to provide Lampert with most of the information he would need.[44]

The Oak Ridge Laboratory expected to complete its study by late summer. The Army would then be able to establish a firm military commitment for its program. In addition, Hertford noted in his analysis to the Chief of Engineers that the AEC would be under new management by then and the Army could expect that "decisions concerning AEC policies in the nuclear field may be somewhat clarified." Furthermore, he explained that the whole question of the future course of nuclear power development is highly controversial. Some factions believed the AEC should demonstrate the feasibility of several types of reactors because of the multimillion dollar investments necessary. Other factions believed the current Atomic Energy Act should be changed to get the government out of the nuclear power development business. In any case, until a national policy was established, Hertford did not think the present AEC would support an Army contract with private industry to develop small nuclear plants for military use.[45]

On his part, following up on the AEC's rejection of the Army's proposal, Gen. Nichols met with Adm. Strauss on May 13 in an attempt to obtain his assistance in reversing the decision. The President's adviser on energy matters suggested that the Army should prepare the studies recommended by the AEC in order to prove more precisely the military's need for atomic power plants. According to Nichols, Strauss' counsel of continued research and patience resulted from the Admiral's knowledge that he would become an AEC commissioner or the Commission's chairman in the near future. He would then be in a position to give the Army's Program a more a favorable reception.[46]

Nichols also reported the situation to Secretary Stevens in a letter dated May 27. Agreeing with Hertford's analysis, he told Stevens that the Commission was undergoing changes in composition at the same time as the Atomic Energy Act was in the process of legislative review and revision. He also explained that the Department of Defense had also "indicated a general unwillingness to support military atomic power projects unless there is a clear-cut military requirement for some." He did, however, cite the Army's progress to date, including the Oak Ridge National Laboratory's study of small nuclear reactors of the type the Army needed, the liaison established between Col. Lampert and the AEC, and Lambert's continuing research into the military's justification of nuclear power plants. Finally, Nichols reported to Stevens on his meeting with Adm. Strauss, saying that the Admiral had indicated that nothing in the Administration's nuclear power policy would prevent the Army's Program from proceeding if it could demonstrate "an economic gain to the use of atomic power in lieu of more conventional power for selected military applications."[47]

Nichols' progress report proved inadequate in satisfying Steven's request for information on the status of the Program. In a Memorandum to the Army Chief of Staff on June 11, 1953, Under Secretary of the Army Earl D. Johnson described Nichols' Memorandum as revealing "very little progress" and claimed he could find no record of being advised of the Army's negotiations with the AEC or the Agency's refusal to authorize the Engineers to proceed with the project. Responding to Johnson on June 18, Nichols suggested that there had apparently been a "considerable misunderstanding that should be clarified."[48]

In attempting to clarify the status of the Program, Nichols did acknowledge that it "had not progressed as rapidly as I would desire." However, he maintained that work had gone forward and Secretary Pace had not indicated any dissatisfaction at his last briefing on January 12, 1953, before leaving office. After summarizing developments since then, Nichols repeated the current status of the Program as described in his May 27 memo to Secretary Sevens. He explained to Johnson that Col. Lampert was following Adm. Strauss' advice to determine whether nuclear power could be economically justified over conventional power for selected military application before again approaching the AEC with a request for support. Nichols also indicated that the Program was "assuming that the justification must essentially depend on this alone and not on the basis that the work we would be doing would be of general value in promoting industrial power." Denying that Secretary Pace had not been kept informed, Nichols concluded that "progress made to date is reasonably consistent with the difficulties encountered in carrying forward military work under the present Atomic Energy Act."[49]

In the meantime, to establish the Army's requirement, Lampert had sought help from the AEC, the Army, and the academic world. He received "a very friendly reception" at the Oak Ridge National Laboratory (ORNL) run by the AEC. Evolving from the Manhattan Project's wartime Clinton Laboratories, ORNL in effect became a technical consultant to the Nuclear Power Division. The Laboratory's director, Dr. Alvin Weinberg, had been one of the early enthusiasts for the development of nuclear power plants for general use; he felt the Army's Program offered considerable promise in helping to develop commercial power. As a result, ORNL became, in effect, a technical consultant to Lampert and the Army's Nuclear Power Program. Working in conjunction with Dr. Hafstad at the AEC Reactor Division, Weinberg agreed to set up a small team within the Laboratory staff to examine the existing nuclear reactor technology and develop some very preliminary designs for suitable power plants. In addition, the team would provide estimates of the plants' characteristics and construction costs.[50]

While the ORNL research progressed during the summer, Lampert received additional inputs from six engineer officers with scientific backgrounds on temporary duty at OCE. They formed a military study group assigned to the Nuclear Power Division to conduct a detailed examination of the military situation around the world in order to determine sites that offered promise as locales for nuclear power plants. The survey focused on power needs in remote areas, such as the Mediterranean and the Middle East, the feasibility of floating nuclear power plants, and the use of mobile power plants for direct support of tactical units. In addition, the Operations Research Office of Johns Hopkins University, working under a contract with the Army, provided assistance to the research effort by setting up a group to do a detailed study of the economic advantages to using nuclear power in remote areas.[51]

While these studies proceeded, Adm. Strauss became Chairman of the AEC on July 2, and on July 13 he asked Gen. Nichols to become General Manager of the Commission. Before deciding to accept the appointment, Nichols went to Gen. Matthew Ridgeway, the Army Chief of Staff, to discuss the offer. He explained that he was happy with his current assignment but would accept the position if Ridgeway thought he could better serve the nation in that capacity. Within the Army, some people thought Nichols should remain in the service and serve as the General Manager while still on active duty. Neither Secretary of Defense Wilson nor the AEC approved this option. Ultimately, President Eisenhower made the decision that Nichols should resign from the Army and serve in the AEC as a civilian. He did remain in his military position throughout the summer, then retiring on October 31, and beginning his new job the next day.[52]

Until he left the Army, Nichols had monitored Lambert's work and, on August 13, he provided Secretary Stevens with an up-date on developments since his June report. Because of the slow progress of Congressional hearings on the development of commercial atomic energy, Nichols wrote that the Atomic Energy Law would not be revised until 1954, with a corresponding delay in any private action to build nuclear power plants.[53]

In assessing the status of the Army's Program, Nichols said research had indicated "that under various conditions there would be a substantial military advantage to the use of nuclear plants in lieu of conventional plants." He noted that capital costs of a nuclear plant could be within a reasonable range of a conventional facility, operating costs would be lower, and savings in logistics would be substantial. He explained that the ORNL investigation had produced a design that could probably be built for $3 million and he estimated that design and construction of a prototype plant would take three to six years. Nichols therefore concluded that the studies done so far justified continuing and expanding the Army's Program with the objective of building a nuclear facility based on the ORNL work.[54]

Nichols then provided the Secretary with a two-phase plan for accomplishing this goal. He said all design work could be done in twelve months at an estimated cost of $500,000. The actual construction would require about two years and cost $2 to 3 million. As soon as the Nuclear Power Division had completed its next report at the end of September, Nichols said he would discuss it with Lewis Strauss and at that point the AEC could determine how much of the actual work it wanted to control vis-à-vis the Army's responsibility.[55]

Following the Secretary's concurrence with the plan on September 1, in his last report before retiring, Nichols provided the Army Chief of Staff with additional information. He said the Nuclear Power Division's Report, now scheduled for completion in October, would recommend that the Army support the design of a prototype nuclear plant that would produce about 1,000 kilowatts of electrical power and sufficient additional steam to heat an arctic installation having a 200-man garrison. According to Nichols, the justification for building such a facility "lies in the fact that 75% of the total logistic support required for such a station would be eliminated by the use of a nuclear power plant." He conceded that this saving would not pay the additional cost of the nuclear plant over a conventional facility in a peacetime situation, but argued that "in time of all-out-war, dollar savings and the assurance of continued operations without dependence on regular deliveries of oil would be an important justification."[56]

## NOTES

Unless specified otherwise, all material cited is located in the Army Nuclear Power Program, Research Collections, Office of History, U.S. Army Corps of Engineers, Fort Belvoir, VA.

1. Public Relations Division, U.S. War Department Release, December 31, 1946, Suid, Box 2.

2. Gordon Thomas and Max Morgan Witts, *Logbook of the Enola Gay,* (Washington, D.C.: National Air and Space Museum, History Office), p. 317; Author's interview with Jacob Beser, Baltimore, MD, Sept. 17, 1985. Mr. Beser served as radar operator on both atomic bomb missions.; Paul Tibbets, *The Tibbets Story* (Long Island, NY: Stein & Day, 1978), pp. 6, 237.

3. Robert Ferrell, ed., "Truman at Potsdam," *American Heritage* (June/July 1980), p. 42; White House Press Release (August 6, 1945), *New York Times,* August 7, 1945, p. 4.

4. Ibid.

5. "Awful Responsibility," *Time,* August 20, 1945, p. 29; Interviews with Gen. Kenneth Nichols, April 24, 1963.

6. Nichols interview, April 24, 1963, Box 8; Gen. K. D. Nichols, to Army Chief of Staff Gen. J. Lawton Collins, July 22, 1952, Box 3.

7. Nichols interview.

8. Ibid.

9. Ibid.

10. Ibid.

11. Ibid.; Secretary of the Army Frank Pace, Jr., Memorandum to Army Chief of Staff Gen. J. Lawton Collins, July 9, 1952, Box 3.

12. Pace to Collins.

13. Ibid.

14. Nichols interview; Gen. K. D. Nichols, *Personal Diary,* July 10, 1952.

15. Gen. K. D. Nichols, Memorandum to Gen. J. Lawton Collins, July 22, 1952.

16. Nichols to Collins.

17. Ibid.

18. Ibid.

19. Ibid.

20. Gen. J. Lawton Collins, Memorandum to the Secretary of the Army Frank Pace, Jr., July 22, 1952, Box 3.

21. Ibid.

22. Ibid.

23. Ibid.

24. Nichols interview; Nichols, *Personal Diary,* July 17, 1952.

25. Secretary of the Army Frank Pace, Jr., Memo to the Army Chief of Staff Gen. J. Lawton Collins, July 29, 1952, Box 3.

26. Ibid.

27. Nichols interview; interview with James B. Lambert, Feb. 16, 1963, Box 7.

28. Ibid.

29. Chief, Special Weapons Branch, Col. F. B. Kane, Research and Development Division, to Chief of Engineers, Lt. Gen. Lewis Pick, Aug. 25, 1952, Box 3; Deputy Chief of Staff for Plans and Research, L. L. Lemnitzer, to Secretary of the Army Frank Jr., Pace, Sept. 4, 1952, Box 3.

30. Secretary of the Army Frank Pace, Jr., Memorandum to Chairman of the Atomic Energy Commission Gordon Dean, Sept. 10, 1952, Box 3.

31. Chairman of the Atomic Energy Commission Gordon Dean, Response to Secretary of the Army Frank Pace, Jr., Oct. 7, 1952, Box 3.

32. Nichols interview; Gen. Kenneth J. Nichols, to Secretary of the Army Frank Pace, Jr., Oct. 16, 1952.

33. Nichols, letter to Pace; Secretary of the Army Frank Pace, Jr., Memorandum to Chairman of the Atomic Energy Commission Gordon Dean, Oct. 20, 1952.

34. Lampert Interview.

35. Ibid.

36. Ibid.; Author's interview with William B. Taylor, Feb. 6, 1963.

37. Nichols interview; Nichols, *Personal Diary,* Dec. 3, 1952.

38. Lampert interview; Taylor interview.

39. Ibid.

40. U.S. Army Nuclear Power Division, *Interim Report, Army Nuclear Power,* Box 3, pp. 34–35; Bill Taylor, interview with author, Dec. 21, 1981, Alexandria, VA, Box 8.

41. Secretary of the Army Robert Stevens, Memorandum to Army Chief of Staff April 13, 1953, Box 3.

42. Nichols interview.

43. Assistant Deputy Chief of Staff Gen. Kenner F. Hertford, Memorandum to Chief of Research and Development Nichols, OCS, April 20, 1953, Box 3.

44. Ibid.

45. Ibid.

46. Nichols' interview; Nichols, *Personal Diary,* May 13, 1953.

47. Gen. K. D. Nichols, Letter to Secretary of the Army Robert Stevens, May 27, 1953, Box 3.

48. Under Secretary of the Army Earl D. Johnson, Memorandum to Army Chief of Staff, June 11, 1953, Box 3; Gen. K. D. Nichols, Response to Under Secretary of the Army Earl D. Johnson, June 18, 1953, Box 3.

49. Gen. K. D. Nichols, Memorandum to the Secretary of the Army, June 18, 1953, Box 3.

50. Lampert interview.

51. Lampert interview, Box 7; Taylor interview, Box 8; interview with Gen. William Gribble, Jan. 23, 1963, Box 7.

52. Nichols interview; Gen. K. D. Nichols, interview with author, April 16, 1982.

53. Gen. K. D. Nichols, Memorandum to Secretary of the Army Robert Stevens, Aug. 13, 1953, Box 3.

54. Ibid.

55. Ibid.

56. Acting Secretary of the Army Earl Johnson, Memorandum to Army Chief of Staff Collins, Sept. 1, 1953, Box 3; Nichols, Memorandum to Army Chief of Staff Collins, Sept. 14, 1953, Box 3.

CHAPTER 2

# ORGANIZATION OF THE NUCLEAR PROGRAM

With the *Second Interim Report,* dated October 30, 1953, Colonel Lampert and the Nuclear Power Division formally established a military justification for the design and construction of a nuclear power plant. The Report concluded that the Army's proposed Nuclear Power Program was compatible with national policy and with commercial interest in the subject. It cited military power and heat requirements that nuclear power plants could meet in remote areas. According to the Report, the small nuclear reactor that ORNL had designed was technically feasible and could be built without delay at comparatively low cost, with a minimum risk. Adapted from Rickover's submarine reactor, the chosen design was a pressurized, water-cooled and water-moderated reactor, fueled with enriched uranium. Although not cheaper than a conventional facility in most locations, the Report argued that a nuclear plant could be justified in certain high fuel cost areas on an overall dollar basis. Moreover, in wartime, a nuclear plant could gain a significant advantage because of logistical problems in supplying conventional power plants.[1]

The Report recommended that the Army build a prototype plant at a site in the United States at an estimated cost of less than $5 million. The initial facility should be quickly followed by a field unit constructed in an arctic location with completion scheduled for late 1958. Although cost of the second unit would depend on the site, the Report estimated $2 to $3 million per year as sufficient funding for the Program. Finally, the Report called for continuing studies to apply advancing nuclear technology to evolving military requirements.[2]

The *Second Interim Report* quickly made its way up the chain of command. On November 10, Col. Lampert briefed the Military Liaison Committee about the Nuclear Power Division's research and conclusions. On November 13, the Army requested that the Defense Department provide an appropriation of $500,000 to initiate the design of the prototype nuclear plant during fiscal year (FY) 1954. The service also recommended to the Joint Chiefs of Staff that it establish a military requirement for the development of such a nuclear reactor. In advising the Military Liaison Committee of its actions on December 4, the Army's Research and Development Office requested that when the Joint Chiefs of Staff approved the requirement, the Committee should forward the *Second Interim Report* to the AEC with the recommendation that it undertake the development program. The Research and Design (R&D) Office also recommended that the Defense Department ask the AEC to undertake the development of the reactor at Commission expense, but explained that if necessary, the Army "is prepared to assist the Commission in funding, or to accomplish the development by direct contract with Commission approval."[3]

As expected, on December 9, the Joint Chiefs of Staff established the requirement for a military nuclear power plant. On December 14, the Military Liaison Committee advised the Atomic Energy Commission of this action. Nevertheless, the problem of how to fund the Program became an issue almost immediately. On December 16, Donald Quarles, the Assistant Secretary of Defense for Research and Development, stated, "We understand that the Atomic Energy Commission is financing the nuclear power plant design project and accordingly we are withholding the requested funds without prejudice to the planned project." But, on February 11, in acknowledging the Liaison Committee's letter, the Atomic Energy Commission indicated it " presently has no funds available" to spend on the project. Instead it suggested the Army should transfer funds to the Commission "to permit initiation of this work without delay."[4]

In anticipation of working out funding problems and in accordance with the Joint Chiefs of Staff action, on January 11, 1954, the Chief of Engineers activated the Nuclear Power Branch at the Engineer Research and Development Laboratories (ERDL) at Fort Belvoir, Virginia. This Branch was to act as the Corps of Engineers field agency for the Nuclear Program. The Branch's first chief, Maj. Joseph Bacci, remained resident engineer for both the Army and the AEC during the construction of the prototype nuclear plant from 1954 to 1957. The Branch itself came directly under the technical supervision of the Nuclear Power Division in the Office of the Chief of Engineers.

Under Col. Lampert, the Division operated as part of the Corps of Engineers Directorate of Research and Development, where it remained until an Army reorganization in 1962. Capt. Bill Taylor served as Lampert's first assistant from October 1952 to September 1953, at which time he contracted a very severe case of polio as a Major, retired from the Army, and was absent until January 1955, when he rejoined the program as a civil engineer. His loss was at least in part offset by the arrival of Lt. Col. William C.

Gribble, whom Lampert described as having provided the Program "tremendous support and assistance."[5]

Col. Gribble had worked in the Los Alamos Scientific Laboratory as a metallurgical engineer from 1948 to 1952, while on loan from the Armed Forces Special Weapons Project. Toward the end of his assignment, he became aware of the Army's developing interest in nuclear power. He also learned that the AEC had given the Army a slot in its new training program in reactor technology located at the Oak Ridge National Laboratory. Discovering that the Army's position had not been filled, Gribble applied through the Chief of Engineers to attend the one-year course at the Oak Ridge School of Reactor Technology and became the first Army officer to enroll in the Program. In January 1953, half way through the course, Gribble met Col. Lampert, who was visiting the ORNL as part of his continuing research. Lampert discussed with Gribble his interest in reactors, the extent of his knowledge of nuclear technology, and asked if he would provide assistance in the development of the Army's Program. As a result, during the balance of Gribble's attendance at the school, he maintained contacts with Lampert and Taylor, for the most part receiving information about the Army's interest in small nuclear plants and helping to screen lists of prospective students from Army personnel applying to the school. Before he had ended his year's study, Gribble received orders assigning him to the Nuclear Power Division.[6]

Gribble's arrival in Washington brought Lampert's staff to four, including his secretary. During the next several months, he helped prepare the *Second Interim Report*. Replacing Taylor as Lampert's right-hand man, Gribble then assisted in implementing the Joint Chiefs of Staff's decision to support the Army's Nuclear Power Program. It received official approval on February 10, 1954, when Secretary of Defense Charles Wilson confirmed the Joint Chiefs of Staff development requirement "for nuclear power plants in supplying power for heat and electricity at remote and relatively inaccessible military installations." Wilson also formally designated the Army as the "cognizant agency" in carrying out the requirement.[7]

By this point, the AEC was finally ready to agree to the Army's original request to establish a joint Army-AEC program as it had done earlier with Adm. Rickover and the Navy. With the help of Dr. Hafstad, the AEC and the Corps of Engineers officially recognized the informal working relationship of the previous 18 months, on April 1, 1954, with the establishment of the Army Reactors Branch in the AEC's Division of Reactor Development. Lampert staffed the Branch primarily with Corps of Engineers personnel and liaison officers from the Air Force, Navy, and other Army technical services. The AEC also allowed Lampert to take initial steps toward establishing a small budget within the Commission, which enabled him to begin recruiting a cadre of civilian technical people on the AEC payroll. At the same time, Lampert worked out arrangements so that his military people were accepted within the AEC as technical members of the Branch.[8]

On the Army side of the organization chart, the Corps of Engineers concurrently established, on April 1, the Army Reactors Group, also with headquarters at the AEC as a focal point for the military side of the Program. For administrative purposes, the Group was attached to the Engineer Center at Fort Belvoir, but reported directly to the Assistant Chief of Engineers for Troop Operations. In fact, it remained a relatively small office with the majority of people in the Army's nuclear program working full time in the AEC.[9]

In practice, the Army Reactors Branch was generally responsible for the research and development of nuclear power plants for land use. Reflecting the increasing importance of the Program, the Chief of Engineers redesignated the Nuclear Power Division, Lampert's initial office, as the Office of Special Assistant for Nuclear Power. The OCE/military element in the dual organization structure became responsible for coordination of Army requirements, dealing with Army and Engineer staff matters relating to nuclear power plants, and planning operational aspects of the Program. Because of his dual role as head of both the Army Reactors Branch and the Office of Special Assistant for Nuclear Power, Lampert was able to provide close and continuous coordination of the program.[10]

Before beginning actual work on the prototype nuclear plant, Lampert still had to resolve the matter of funding. In advising the Secretary of the Army of the AEC's position on providing money for the military Program, Maj. Gen. John P. Uncles, the Army's new Chief of Research and Development, explained that the service felt it "highly desirable that this project be contracted on a lump sum basis inasmuch as there had been a strong indication that potential contractors may be willing to undertake this effort at cost or less than cost in order to get established in this field." However, he advised that the service's Nuclear Power Program had never expected a financial commitment of such magnitude in FY 1954 "even though it has been recognized as an important long range development which the Army should certainly support." Gen. Uncles noted that the Program was, in fact, an all-military requirement, that the *Second Interim Report* was completed by the time FY 1954 budget-planning had been finalized, and that the AEC should normally be expected to fund a nuclear program such as the Army's. Consequently, he advised Secretary Stevens that "justification existed for requesting that the Secretary of Defense's emergency funds be made available for financing the program."[11]

In response, Stevens sent a Memo on March 4 to Assistant Secretary of Defense (Research and Development) Donald Quarles outlining the current situation. He pointed out that contrary to Quarles' assumption in his December 16, 1953 Memo, the AEC did not have funds available for the Army's Nuclear Program. Nevertheless, given the recognition of the importance of developing the military's nuclear power capability, Stevens stated his belief that the AEC should not be delayed in negotiating a contract for building the initial power plant. The Secretary acknowledged that whereas the Army could probably rework its FY 1955 and FY 1956 budgets to come up with the required $4 million to finance the prototype plant, he said this would have a significant impact on the service research and development program. Consequently, Stevens asked Quarles to reconsider his December decision to disapprove the Army's request for emergency funds from the Department of Defense (DOD) research and development allocations. In support of his request for the $4 million he cited the overall military requirement for nuclear power and the AEC's normal role in providing such funding.[12]

At this point, Kenneth Nichols reinvolved himself with the Army's Program by discussing the funding issue with Adm. Strauss. They agreed that the prototype nuclear reactor should represent a step in development rather than the production of a fully developed item. At the same time, they agreed that the AEC had the responsibility of carrying the development of the reactor portion of such a power plant through the prototype stage. To fund the project, however, required a new line item in the AEC budget; Nichols asked both the appropriate House and Senate Committees to instruct the AEC to insert the item in its budget.[13]

Following up his discussion with Strauss, Nichols called Quarles on March 12 to brief the Secretary on the results of the meeting. They then resolved to wait for Congressional action before proceeding any further because the AEC could not act without such a line-item appropriation. In addition, they also agreed that the AEC and the Defense Department should, if possible, work together to find a budget approach to keep the Army's Nuclear Program on the track.[14]

On his part, Quarles followed up the conversation by informing Secretary Stevens later in the day that he thought the Army should proceed on the assumption that the project would continue with the service funding the conventional power equipment and the AEC providing money for the nuclear portion of the plant. If this arrangement did not prove practical, Quarles said he would work with the Army to find an alternative source of funding for the program.[15]

The Defense Department and the AEC then proceeded in the following months to find an acceptable formula for financing the Army's Program. On July 8, Quarles formally asked Strauss and the AEC to collaborate with the Department of the Army in developing the Army's Nuclear Program. He acknowledged the advantages to the military of a transportable nuclear power facility and of having a prototype available as soon as possible for training military personnel in its operation. To Quarles, however, "the primary objective of the program should be to make not only a valuable, but the largest possible, development step toward a practical nuclear power unit for Army use in the field." In accomplishing this, he said the Defense Department would rely on the AEC "to determine what reactor development, meeting the above criteria, is most suitable for this application." In regard to financing the project, Quarles said the Defense Department "is prepared to share equally with the AEC the estimated 4 million dollar cost of the first complete prototype power unit."[16]

Acting on this proposal on July 21, the AEC approved the joint funding of a lump sum contract "for the detailed design, construction and test operation of the Army Package Power Reactor (APPR)." The AEC agreed to pay for the cost of the reactor and associated equipment, with the Army paying for the cost of the conventional portions of the plant. In its confirmation letter to the Military Liaison Committee on August 19, the AEC said that it would soon be extending invitations for proposals to qualified contractors and expected that the AEC would award a contract approximately four months from the mailing date of the bid announcement.[17]

With funding and responsibility arrangements resolved, the Army's Research and Development Office advised the Chief of Engineers on September 9 to proceed with the Nuclear Power Program along the lines of agreement between the AEC and the Defense Department. The Office also approved the decision already made by Lampert and Dr. Hafstad in the AEC's Reactor Development office to use the light water pressurized reactor in its prototype facility. In addition, it asked that the Nuclear Program "give attention to the possible application of nuclear power to meet other power requirements of the Army than those of fixed land installations." According to the Office, these potential uses of nuclear power included land vehicles, railroad locomotives, and floating power plants. When the technology had advanced sufficiently, the Office wanted to know which application should be developed.[18]

Before such advanced activities could even be considered, of course, Lampert had to get to work on the prototype plant. Apart from the decision to use a light water reactor similar to the one the Navy had developed for its nuclear submarines, Lampert and his staff had to settle on a location for the first Army reactor. At first, they considered building the prototype plant in a remote location to ascertain if the concept of small nuclear plants operating under difficult conditions was valid. Ultimately, the Nuclear Power Group concluded that the prototype plant simply had to be built in the United States.[19]

According to Lampert, his staff "just didn't feel that the technology was far enough along so that we could go immediately to some place like Greenland or the northern part of Alaska and start building a plant and have a reasonable assurance of getting it finished without extreme delays and lots of trouble." At the same time, the Group realized it would need a training facility to teach military personnel how to operate the several power plants the Army expected to build. Consequently, it evolved the idea of building the prototype plant in the United States and having it serve as both an experimental and a training facility.[20]

Where to actually build the plant remained the subject of much discussion. To solve the problem, Lampert set up a joint board of representatives from the AEC and the Corps of Engineers to study possible site locations belonging both to the Army and the Commission. Using a set of criteria it had drawn up, the board decided on June 29, that the prototype plant should be built at Fort Belvoir. Despite the careful selection process, concern was expressed about the choice, especially from the Technical Advisory Panel on Atomic Energy (TAPAE) in the DOD Office of Research and Development. In the Panel's fifth meeting held on June 14, 1954, it had questioned the advisability of installing the first reactor at Belvoir "where presumably there was no existing facilities for handling the emergencies that might arise in building and testing a new type of reactor."[21]

In response to the Panel's concern, Maj. Gen. R. F. Hertford, now Chief of the Army's Research and Development Office, acknowledged the lack of facilities relating specifically to nuclear reactor emergencies. He then explained that the Corps of Engineers did have its Engineer Research and Development Laboratories (ERDL) at Fort Belvoir. While the ERDL were not primarily involved with atomic energy work, the Research and Development Office pointed out that the Laboratories

> are extensively concerned with various complicated and delicate programs, including work with industrial gases, petroleum products, and explosives. The laboratory includes groups working on sanitary engineering and water purification problems involving radioactive contamination, and on radiation effects on various materials. Radiation sources are used under AEC approved procedures and proper health physics practices are followed.

Consequently, the Office concluded that the base did "have existing facilities peculiarly well adapted to control emergencies which might arise, and this was one factor which entered into the selection of Fort Belvoir as the site for the first reactor."[22]

With the funding issue resolved and the site selection made, Lampert turned to the actual building of the prototype plant. Using the preliminary design drawn up with the assistance of the Research Group at the ORNL, Lampert and his staff developed specifications based on the plant's broad objectives. On the military side, these included: (a) solving technical construction and operation problems associated with a reliable nuclear power plant; (b) providing firm cost information, operating parameters, and engineering test data necessary to adapt the system to a remote arctic location; and (c) providing a training facility for troops and specialists who might eventually be required to operate and service remote plants.[23]

At the same time, Lampert took into consideration the needs of the AEC's development program. These included fuel plate and control rod development, radiation effects on fuel plates, critical mass experiments, control rod drive mechanism testing, physics calculations, and simulation tests. The heating requirements at Fort Belvoir were small, thus the specifications eliminated the need for the plant to produce steam heat as the field facilities would be expected to do. Finally, the specifications called for the potential for air transportability and ease of construction. To accomplish these criteria, no component except the turbo-generator and main condenser were to weigh more than 10 tons and have dimensions to exceed 7′ × 7′ × 18′ with maximum weight of the two other components limited to 20 tons. Field erection at a remote site was to require no more than six months.[24]

To fulfill these specifications, the Army Reactors Branch decided to use a lump-sum, fixed-price contract. The Technical Advisory Panel on Atomic Energy, in its June 14, 1954 meeting, also expressed some misgivings over the type of contract to be used. It suggested that a fixed-price arrangement "might make it difficult to incorporate desirable developments as the construction proceeded." Nevertheless, on August 20, the Reactors Branch sent out invitations to thirty-three major construction firms to submit fixed-price bids for the engineering design, construction, and test operation of the prototype nuclear plant.[25]

In explaining his decision to use the fixed-bid procedure, Col. Lampert advised Gen. K. F. Hertford, Chief of the Army's Office of Research and Development: "Industrial enthusiasm in atomic energy permits application of competitive bidding and fixed price contracts to research and development without sacrifice of technical objectives." In fact, eighteen firms submitted bids. Lampert and his staff, as well as a technical advisory group from the original Oak Ridge panel, then subjected the proposals to "a very intensive technical examination and comparison" for more than a week. Ultimately, on December 10, 1954, the Army Reactor Branch completed negotiations for the contract with the American Locomotive Company of Schenectady, New York.[26]

When he advised the Coordinating Committee on Atomic Energy of the Army's decision to use the fixed-bid procedure, Hertford acknowledged that this type of contract rarely was used for a research and development project because of the potential difficulties of incorporating new technical developments into the blueprint after construction began. To handle this situation, however, the contract called for ALCO Products (American Locomotive's new name) to institute "a substantial research and development program" during construction with "important supporting research and development by the AEC." Given this stipulation, Hertford said that both the Army and the AEC anticipated "that there may be desirable developments which should be incorporated as construction proceeds, and are prepared to negotiate change orders to the basic contract plus any change orders would still be significantly less than the cost to the government under a cost-plus-fixed-fee contract."[27]

The actual contract provided a fixed-price of $2,096,753 and included provisions for guaranteed operation to be proven during a 700-hour continuous operational test, followed by a 6-month test period during which the plant would produce usable power. The contract set the completion date as December 10, 1957, exclusive of the 6-month test period. To the Army, the operating plant would determine that the design "will meet urgent military requirements for small capacity nuclear power plants and would provide engineering data and experience which will be extremely valuable in the later programs to develop mobile plants or plants capable of disassembly for removal or relocation."[28]

## NOTES

1. U. S. Army Nuclear Power Program, *Second Interim Report,* Nuclear Power Division, U.S. Army, Oct. 30, 1953, Box 3.
2. Ibid.
3. Army Research and Development Office, Memorandum to Secretary of the Army, Feb. 26, 1954, Box 3; Army Research and Development Office, Memorandum to Military Liaison Committee, Dec. 4, 1953, Box 3.
4. Army Research and Development Office, Memorandum to Secretary of the Army, Feb. 26, 1954; Atomic Energy Commission Chairman Lewis Strauss, Letter to Military Liaison Committee, Feb. 11, 1954.
5. Interview with Col. James Lampert, Feb. 16, 1963.
6. Interview with Brig. Gen. William C. Gribble, Jan. 23, 1963.
7. Secretary of Defense Charles Wilson, to the Chairman, Joint Chiefs of Staff, Feb. 10, 1954.
8. Lampert interview; U.S. Army Nuclear Power Program, *Progress Report,* Office of the Chief of Engineers, U.S. Army, June 30, 1956 (cited hereafter as *Progress Report*).
9. Ibid.; Department of Army, OCE, *General Order No. 3,* April 9, 1954.
10. Lampert interview; *Progress Report.*
11. Army Chief of Research and Development Maj. Gen. John P. Uncles, to Secretary of the Army Robert Stevens, Feb. 26, 1954.
12. Secretary of the Army Robert Stevens, Memo to Assistant Secretary of Defense Donald Quarles, March 4, 1954.
13. Nichols interview; Assistant Secretary of Defense Donald Quarles, Memorandum to Chairman Robert LeBaron, Military Liaison Committee, March 12, 1954.
14. Quarles, Memorandum to LeBaron.
15. Assistant Secretary of Defense Donald Quarles, Memorandum to Secretary of the Army Robert Stevens, March 12, 1954.
16. Assistant Secretary of Defense Donald Quarles, Letter to Chairman of AEC, to Lewis Strauss, July 8, 1954.

17. U.S. Atomic Energy Commission Acting Chairman H. D. Smyth, Letter to Gen. Herbet B. Loper, to the Military Liaison Committee, Aug. 19, 1954, Box 3.

18. Army Research and Development Office, to Chief of Engineers, Sept. 9, 1954.

19. Lampert interview.

20. Ibid.

21. Lampert interview; Technical Advisory Panel on Atomic Energy, to Chairman, Coordinating Committee on Atomic Energy, Feb. 10, 1955.

22. Army Research and Development Office, to Chairman, Coordinating Committee on Atomic Energy, April 12, 1955.

23. Lampert interview.

24. *Progress Report,* pp. 26–27.

25. Lampert interview; Technical Advisory Panel on Atomic Energy, to Chairman, Coordinating Committee on Atomic Energy, Feb. 10, 1955; *Progress Report,* pp. 27–28.

26. Col. J. B. Lampert, Memorandum to Gen. K. F. Hertford, March 11, 1955; *Progress Report.*

27. Army Research and Development Office, Memorandum to Chairman, Coordinating Committee on Atomic Energy, April 12, 1955, Box 3.

28. Ibid.; *Progress Report.*

CHAPTER 3

# BUILDING THE PROTOTYPE

Coincidental with the final negotiations for the contract to build the prototype plant at Fort Belvoir, the Atomic Energy Commission and the Department of Defense reached an agreement on a "Statement of Policy for Operation of Military Power Reactors." Initially, the AEC had forwarded to the Defense Department a draft proposal of the Statement on March 1, 1954. During the following months, the draft underwent a series of reviews and minor revisions until the final version was approved on November 8, 1954. It not only delineated the respective roles of the military and the AEC, but assigned responsibility within the Defense Department to each service for operating its own reactors.[1]

The AEC was to transfer required quantities of fissionable material to the military as needed by each reactor. The Commission's responsibilities were to "end at the moment of transfer except that the Atomic Energy Commission will, as requested by the military department concerned, continue to review and evaluate operating procedures, general safety standards, and the Department of Defense security program related to the handling and use of fissionable materials." In turn, following the transfer, the service assumed full responsibility for safeguarding the fissionable material and security for any classified information related to the reactor. At the same time, the AEC retained the responsibility for "determination of what constitutes Restricted Data."[2]

Each military service was to establish and observe safety standards relating to the health and hazards to life and property and assume payment for any damages to life and property resulting from the use of fissionable material. The service was also to provide the AEC with all pertinent data and information concerning the military's reactors "including safety standards and any changes thereof, and operational experiences particularly with regard to significant reactor malfunctions." Except for the reactor programs of a developmental nature, each service was to fund the cost of fuel fabrication.[3]

In turn, each service was to return to the AEC the spent fuel elements for recovery of fissionable material "at a time and place and in a physical form mutually agreeable to the Atomic Energy Commission and the military department concerned." The AEC did expect, however, that the services should pay for both the costs of recovery and the fissionable material lost in processing. The AEC agreed to make its facilities available to the services for both fabrication and recovery of the material as long as this was the most advantageous method of operation. Through discussions, the AEC and the military agreed to reach a determination as to the reprocessing costs. Likewise, through mutual agreement, the services and the AEC would determine the amount of material consumed or lost in each military reactor with the services paying for the cost of the fuel subsequent to FY 1955.[4]

With the guidelines in place and the contract signed, the Army and ALCO could concentrate on building the prototype plant, initially called the APPR–1 (Army Package Power Reactor #1) and subsequent to the adoption of the standard nomenclature in 1958, designated the SM–1 (Stationary Medium Power Prototype). ALCO came to the nuclear power plant field with a background as a heavy equipment manufacturer that had supplied equipment to atomic energy installations for 10 years. In turn, the company selected Stone and Webster, the principal contractor on the Manhattan Project, as architect and construction subcontractor with responsibility to build the conventional or steam portions of the plant.[5]

Although the AEC's Chicago Operations Office had served as contracting agent for the project, the Commission's Schenectady office took over supervision of the contract because of its proximity to ALCO's headquarters. In turn, the Schenectady office designated Maj. Joseph Bacci, Chief of the Nuclear Power Branch at Fort Belvoir, as the site resident engineer with responsibility for day-to-day inspection and check out of construction activities.

The Army Reactors Branch and the Package Reactor Group at ORNL closely monitored ALCO's design work through the AEC's approval of the plans. The facility, on which construction began October 5, 1955, incorporated many innovations that represented state-of-the-art development of a nuclear power plant anywhere in the world. Externally, the vapor containment shell became the most visible feature of the plant. In light of the SM–1's location within 18 miles of the White House, the plant's designers determined very early that the reactor should have maximum containment against any possible nuclear accident that might release radioactivity into the environment.

To accomplish this, ALCO designed the vapor container to enclose the entire reactor primary system. As a result, a cylindrical steel shell with hemispherical top and bottom caps enclosed the reactor vessel, the primary cooling pumps, the heat exchanger, and the pressurizer. The inner shell ensured that any spills could rapidly be cleaned off the steel surface. The cylindrical containment vessel was surrounded by a 2-foot layer of concrete, which in turn was surrounded by a second steel shell. The planners intended that the concrete layer would stop any fragments from a possible steam explosion inside the containment vessel. The outer shell formed the actual containment vessel and was designed to keep any gas or other fission products within the shell.

Given this construction, only tolerable amounts of radioactive material could possibly reach the atmosphere in the event of an accident and would not pose any danger to people or the environment. Nevertheless, such built-in safeguards did not prevent residents across the Potomac from claiming damage to their property from nuclear radiation—even before the plant received its first shipment of uranium fuel. One man blamed the reactor for causing paint to peel from his house and a woman blamed radioactivity from the still-inactive plant for the sickly condition of her roses. To contradict such claims, Bacci planted his own roses within the restricted area of the power plant and at least one rose bush remained alive 8 years after the SM–1 was deactivated after 16 years of accident-free operation.[6]

Within the reactor, stainless steel was used for the first time as a cladding material for the flat-plate fuel element. Many nuclear engineers questioned the choice of stainless steel in an atomic reactor, believing it would absorb too many neutrons, in light of the high neutron cross-section of the constituent elements of stainless steel. Some of the doubters even suggested that the reactor might never go critical because it would contain too much poison material. Despite the criticism, the metallurgical engineers at ORNL developed and fabricated the fuel element that was to be used in all succeeding Army reactors. Moreover, the nuclear power industry ultimately adopted stainless steel as the fuel element cladding material for commercial power plants, rather than the more costly zirconium alloys used in naval pressurized water reactors.

The fuel element used a mixture of uranium dioxide ($UO_2$) and stainless steel. The mixture was compacted and clad with plates of stainless steel using the "picture frame technique." The " picture frame" was then hot-rolled to the correct length, width, and thickness. Eight of these plates were then brazed into " side-plates" made of stainless steel with endboxes attached to them, forming a single fuel element. The SM–1 reactor had 38 such fuel elements in its core to constitute the critical mass.

To control the reaction, the reactor was to have 7 control rods. As an additional control mechanism, the SM–1 used boron carbide as a burnable poison incorporated into the fuel leaf of each of the fuel elements. The inclusion of the boron carbide allowed the reactor to go critical and permitted the complete control of the core throughout its life. At the same time, it extended the life of the initial core from its original design of 15 megawatt years to about 16½ years. The success of the innovation caused several civilian and military reactors to adopt the concept of using burnable poison as a means of reactor control.

The AEC conducted the research and development work on the core and other aspects of the reactor at the Oak Ridge National Laboratory under the guidance of the Package Power Group, the technical advisory task force established in FY 1954 under ORNL's Electro-nuclear Research Division to study small nuclear power plants. In addition to the work on the core, the Advisory Group conducted a series of experiments to determine critical loading, control rod worth, as well as power and flux distribution during the period from October 1955 to February 1956. In correcting factors previously omitted or erroneously evaluated in calculated values, the Group obtained a more accurate determination of the critical mass, decided that the SM–1 should have seven control rods instead of the five called for in the original design, and emphasized certain heterogeneity effects of power and flux distribution. During the five months of research, five or six people worked full-time on the experiments.

Meanwhile, demonstrating its desire to proceed as rapidly as possible with the construction of the SM–1, the Army had negotiated a modification of the original contract almost immediately after the groundbreaking of the power plant. On October 25, 1955, ALCO agreed to the acceleration of construction at the cost of an additional $223,500 to the AEC. By March 1, 1956, the company had completed approximately 40 percent of the work toward the new completion date of July 10, 1957.[7]

In its eagerness to expand involvement in the nuclear field, the Army had begun an evaluation of a variety of possible military uses for atomic power, even before the design work had been completed on the SM–1. On February 4, 1955, the Chief of the Army Communications Service requested that the Chief of Engineers provide information "on the feasibility of utilizing nuclear power plants" at Army Command and Administrative Network (ACAN) station sites around the world where the Army Signal Corps intended to install communications equipment. In response, the Engineers Office described the features of the prototype plant and explained that the project had among its purposes "to determine the reliability of a nuclear powered electrical generating facility. Estimates of potential fuel logistic savings indicate that a nuclear system will be essentially independent of supply lines." The Memorandum also stated that the Signal Corps facility at Wildwood Station in Alaska "is a logical choice as a site for the first operational nuclear power plant" and indicated that the Nuclear Power Program would proceed with appropriate studies if the Signal Corps concurred.[8]

In a similar search for potential applications of nuclear energy, Col. Gribble conducted a briefing for the Transportation Corps on March 15, 1955, regarding possible uses for nuclear power by that organization. In particular, he discussed the general problems surrounding the development of a nuclear-powered snow train and stated that such an application "appears to be quite feasible." He suggested conducting a feasibility study using nuclear power to propel the train to compare it with diesel-electric power, then being developed in the Transportation Corps' research program. As a result, the Chief of Transportation decided to " officially inform" the Chief of Engineers that the Transportation Corps had an "interest in a Nuclear-Powered Snow Train and/or Beach Lighter and request Chief of Engineers to adapt the Nuclear Power Program to include assistance to the TC in these fields."[9]

The Army's interest in small-scale applications of atomic energy received an added boost in August when Donald Quarles, Assistant Secretary of Defense for Research and Development, reiterated to Lewis Strauss the military's need "for an over-all program aimed at development of the smallest family of nuclear power plants capable of satisfying the basic requirements of all the military services for heat and electric power under the adverse conditions encountered at remote locations." In his letter to the AEC Chairman, Quarles explained that the three armed services "now emphasize the increasing importance of small plants in the low end of the power range of interest to the DOD." Citing the AEC's own recently completed feasibility analyses, the Assistant Secretary advised the AEC that such small reactors should have output capacities approximately one-tenth that of the SM–1 and "should be designed for minimum maintenance and practically unattended operation. They should be air transportable to remote sites for relatively easy erection in a short construction season."[10]

Quarles also stated the Defense Department's desire to encourage the AEC "to emphasize in its current program research and development leading to early availability of nuclear plants with reactors in the 1 to 2 megawatt range, for supplying about 200 kilowatts of electrical power and the heat necessary for operation in remote locations. When this research and development has reached the proper stage and satisfactory designs are available, the DOD will consider any action necessary to establish a specific prototype development project." Apart from this requirement, the Pentagon stated its desire "to encourage the Commission to continue a vigorous program of examining all feasible reactor concepts for application to military requirements in remote areas."[11]

Quarles' letter initiated the AEC's study of the military requirements for portable nuclear power plants. By the summer of 1956, the investigation indicated "the desirability of actively continuing the current project for development of gas-cooled reactors." Consequently, the Office of Chief of Engineers advised the Army's Research and Development Office that "it is considered timely to encourage the Commission to pursue this program actively in order to achieve early availability of experimental data required for definitive prototype designs."[12]

In the meantime, work had progressed on the SM–1 without significant problems and ALCO completed construction in 18 months. The plant itself was a pressurized water reactor, rated at 10 megawatts of heat, and used highly enriched uranium as its fuel source. The power cycle was divided into two systems. The primary system employed a nuclear reactor instead of a conventional boiler to heat water to high temperatures under high pressure. The secondary system closely approximated the power-generating equipment of a conventional power plant using the high temperature pressurized water to generate steam for the steam turbine that turned a conventional 2,000-kilowatt electrical generator.

The primary system included the reactor vessel and its core of 38 fuel elements and 7 control rods, 2 primary coolant circulating pumps, the steam generator, the pressurizer, a water purification system, and necessary piping and shielding. Primary system cooling water entered the reactor vessel at 431°F under a pressure of 1,200 pounds per square inch (psi), which prevented the water from boiling. As the water circulated upward between the fuel plates at a rate of 4,000 gallons per minute, the reactor heated it to temperatures of up to 450° F. From there, it passed to the steam generator. The generator, a shell-and-tube heat exchanger, received the reactor-heated water that passed its heat through tube-walls to the secondary system water that was converted to steam. The primary cycle was completed as the water from the steam generator was returned to the reactor vessel by means of either of two special canned rotor pumps, designed to permit absolutely no leakage.

The SM–1's secondary system consisted of the shell (secondary) side of the steam generator, the turbine-generator set, the main condenser, the feedwater heater, the evaporator, and necessary pumps, storage tanks, and interconnecting piping. The secondary system water did not come in contact with any direct radiation, consequently the secondary system components required no shielding and except for the steam generator, the entire system was located outside the vapor container. The steam produced in the steam generator drove an eight-stage turbine-generator and was then exhausted to a surface de-aerating condenser where water from the Potomac River cooled the steam back to its liquid form. Pumps then circulated this water from the condenser hot well through a feedwater heater and then back to the steam generator. To minimize corrosion in both systems, the plant provided water purification and treatment, which maintained the total solid concentration in the primary water at less than two parts per million, with the secondary water only slightly less pure.

The shielding of the reactor and its associated equipment was divided into two sections. A 13-foot-diameter steel tank filled with water and containing eight concentric steel cylinders, each 2 inches thick and separated from each other by a 1-inch layer of water formed the primary shield. The designers elected this iron-water shield because it was smaller, lighter, and easier to assemble than a concrete shield of equal protective power. The vapor container, consisting of concrete and 7/8-inch-thick steel plate reduced the external radiation release into the atmosphere well below the laboratory tolerance level mandated by the AEC.

The SM–1 produced its electric power from a steam turbine-driven 2,500 Kilovolt-ampere (kva) generator, which operated (through a reduction gear) at 1,200 revolutions per minute (rpm) and produced three-phase, 60-cycle current at 4,160 volts. To operate the plant's pumps, controls, and utilities, approximately 180 kilowatts of power was tapped off, with the remainder of the plant's output to be fed into the Fort Belvoir electrical net in parallel with the local private utility company. The output circuitry was so arranged that the plant could supply much of the base's electric power requirement while still not feeding power into the commercial utility grid.

The power level of the SM–1 was regulated by a combination of manual and automatic control systems. Pressurized water reactors have an inherent control characteristic known as "negative temperature coefficient." As a result, when the water temperature in the reactor increases, the "reactivity" of the core decreases and vice versa. A decrease in plant load results in less steam being withdrawn from the steam generator, which causes the water temperature in the reactor to rise. The negative temperature coefficient then reduces the reactor power level, compensating for the reduced plant output.

Apart from the inherent control of the reactor provided by the negative temperature coefficient, the SM–1 had electrical and mechanical control circuits that made a nuclear accident virtually impossible. At the same time, the plant's operator console did contain a large red button labeled "SCAM." If activated, the button dropped the control rods into the reactor core and the reactor shut down within a fraction of a second. The controls for regulating the electrical-generating machinery were similar to those used in a conventional steam power installation.

In operating the SM–1 and subsequent field nuclear plants, Col. Lampert faced the problem of dealing with a new technology and lack of personnel trained in reactor operations. Recognizing the need for highly skilled technicians and supervisors who would be able to draw on their own resources in remote areas far from immediate assistance, Lampert gave a great deal of attention to the training aspects of the Nuclear Program. Early in 1956, he selected Maj. Ernest Graves, Jr. to develop the operator training program. In setting up his school, Graves decided that the first class would form the cadre for training all subsequent power plant crews.[13]

Although various universities and scientific institutions offered advanced courses in nuclear physics, Graves organized his own program for training plant operators at Fort Belvoir. He selected the cream of the warrant officers and enlisted men who began applying to the training program in recognition of its unique status and the opportunities it offered. Applicants were limited to career soldiers who volunteered for the training and whose aptitude test scores at least equalled the minimum standard for admittance to officer candidate school. The selection board carefully screened each applicant, taking into consideration such factors as civil and military education, aptitude, reliability, and maturity. The officers in the program came from the Army Corps of Engineers and Signal Corps as well as from the other branches of the armed forces. They were expected to have a master's degree in nuclear engineering or a closely related scientific discipline.

In May 1956, the men chosen to form the first military crew of the SM–1 began a five-month course at the University of Virginia in math, physics, and mechnical engineering. Five commissioned officers attended two- and four-month nuclear engineering courses at Pennsylvania State University, North Carolina State University, and the University of California, Berkeley. The instrumentation group (an officer and four non-commissioned officers from the Signal Corps) received training in instrument maintenance at Minneapolis-Honeywell Regulator Company, the Brown Instrument Company, and the Foxboro Instrument Company. Upon completion of these courses, the men underwent on-the-job training at the ALCO plant in Schenectady. Then, they reported to Fort Belvoir during the final stages of the SM–1 construction to participate with the contractor's technicians and engineers in the final testing and start-up operation.

On April 8, 1957, the SM–1 was brought to its initial criticality when the control rods were withdrawn until a self-sustaining chain reaction began in the reactor. Between April 8 and April 15, the crew ran a series of final checkout operations. On April 15, the SM–1 generated electric power for the first time. However, on April 18, with the formal dedication only 11 days away, the power plant building began to vibrate when the electric generator reached operating speed. As the crew discovered, the problem was in the rotor of the standard, off-the-shelf General Electric generator, not in any of the new nuclear components. Col. Lampert called Ralph Cordiner, President of General Electric in Schenectady, to arrange for a special run of copper wire needed to rewind the rotor.

Although the SM–1 crew disassembled the rotor and shipped it to the General Electric plant in Baltimore by April 22, the agreed-upon time, the wire had not arrived. Military police and state troopers from New York, New Jersey, Pennsylvania, and Maryland were immediately mobilized to find the truck. Within an hour, the police located the truck in Baltimore making an unscheduled delivery. Once the load of copper wire arrived at the plant, General Electric technicians rewound the rotor in an around-the-clock operation. Then, they applied a protective coating of waterproof materials to the rotor and prepared it for shipment in late afternoon on April 26.

The military convoy, consisting of the truck, a spare truck in case of a breakdown, and motorcycle police with sirens wailing, made it to Belvoir by late in the evening. The crew began work immediately in an effort to get the plant back in operation for the formal dedication on April 29. Although the crews completed the work on April 28, Lampert and his men had no assurance that the SM–1 would produce electricity for the assembled guests. Some of the staff even suggested that if the plant could not be brought back on-line by the time of the scheduled ceremony, they could use electricity from the regular commercial grid to light the incandescent bulb designated to signify the dedication of the facility. Lampert vetoed the idea immediately, saying there would be no light unless the SM–1 generated the electricity. In the end, with the rotor back in place, the plant crew brought the reactor back to criticality, and the SM–1 began generating power early on the morning of April 29.[14]

Col. Lampert had originally rejected the idea of a dedication by enthusiastic Army officials as soon as construction had begun. With the plant nearing completion, however, the Office of the Chief of Engineers set about to inaugurate the facility in style with a joint Army–Atomic Energy Commission ceremony on April 29. Initially, the Engineers had hoped President Eisenhower would give the dedication speech, followed by the "initial application of power to an appropriate military use, viz, military equipment such as radar, and possibly to power television equipment." The planners proposed that if the President could not attend the ceremony, Secretary of the Army Wilber Brucker and Adm. Lewis Strauss deliver the principle addresses and "together they throw the switch to apply nuclear power to an appropriate military use." The Army expected to invite 300 people from both government and private agencies and ALCO was given permission to host the postdedication reception.[15]

Pursuant to these plans, on behalf of the Army and the AEC, on February 8, Secretary Brucker invited President Eisenhower to attend the ceremony. In his letter, Brucker explained that the SM–1 "will be the first military nuclear power plant built for the generation of electrical power." He pointed out that in addition to meeting military requirements, the plant "also offers the possibility of meeting certain civilian needs for small blocks of power and for export to foreign countries."[16]

Ultimately, however, President Eisenhower did not attend the dedication, leaving Brucker and Strauss to throw the switch. To symbolize the military and civilian promises of nuclear power the electricity generated during the ceremony powered a radar antenna and a printing press. Secretary Brucker read President Eisenhower's congratulatory message and in his own prepared remarks spoke about the revolutionary potential of nuclear energy. In his address, Adm. Strauss observed that the armed forces now were in a position to possess "compact, efficient, and readily transportable power plants for remote outposts where electrical energy would be unattainable, except under conditions of extreme logistical strain." He stated that the SM–1 "is as surely a part of President Eisenhower's concept of 'Atoms for Peace' as are the isotopes from our atomic reactors which are today combatting man's diseases and suffering, improving the fertility of our fields and the products of our factories and pointing the way to new discoveries."[17]

With the dedication completed, the SM–1 embarked upon the required period of testing stipulated in the Government's contract with ALCO. Many nuclear experts had believed that the 700-hour performance test would require several attempts before it could be completed with no more than the 40-hour limit of total "down-time" mandated in the contract. In fact, the reactor initially behaved very closely to its theoretical predictions and, after minor adjustments and preliminary tests had been completed, the SM–1 began its 700-hour test on June 2. Operating at designated loads varying between 10 percent and 100 percent of design capacity, the plant was shut down for a total of only 7 hours and 28 minutes for special instrument tests during the test period. Its performance exceeded that which the designers and builders had predicted. In particular, the reactor demonstrated the inherent control characteristics resulting from the negative temperature coefficient as it responded to rapid changes in electrical load with speed and stability. After successfully completing the 700-hour test on July 2, the SM–1 was shut down for adjustment and the correction of minor faults in the secondary system.

The plant then underwent a 6-month sustained operating test, which required it to maintain full power for twenty-one days and at power loads varying from zero to full power for the balance of the period. During this test, the military technicians continued their on-the-job training. The plant's few unplanned shutdowns resulted from the failure of off-the-shelf conventional equipment, not from any breakdowns in the primary reactor system. At the completion of the 6-month test on January 24, 1958, the government accepted the plant from ALCO. The AEC then directed its operation until July 1, 1960, when it transferred control to the Chief of Engineers.

## NOTES

1. *Statement of Policy for Operation of Military Power Reactors*, Nov. 8, 1954, Box 3.
2. Ibid.
3. Ibid.
4. Ibid.
5. In addition to individual documents cited, the discussion of the building of the SM–1 comes from William B. Taylor, "The Army Package Power Reactor," *The Military Engineer*, no. 332 (November-December 1957): 425–429; and "SM–1 Nuclear Power Plant" and "Technical Information, Army Package Power Reactor," ANPP, Research Collections, Office of History, U.S. Army Corps of Engineers, Box 6.
6. The author personally saw the bush in bloom in 1981.
7. Army Chief of Staff, *Memorandum for the Record*, March 14, 1956, Box 3.
8. Chief, Army Communications Service Division, to the Chief of Engineers,

Feb. 4, 1955, Box 3; Assistant Chief of Engineers for Troop Operations, to Chief Signal Officer, April1, 1955, Box 3.

9. Army Corps of Engineers, *Memorandum for the Record*, March 16, 1955, Box 3.

10. Assistant Secretary of Defense Donald Quarles, letter to Chairman Atomic Energy Committee Lewis Strauss, Aug. 3, 1955, Box 3.

11. Ibid.

12. Assistant Chief of Engineers for Troop Operations, to the Army Research and Development Office, June 29, 1956, Box 3.

13. Interview with Col. James Lampert, February 16, 1963.

14. Interview with Maj. Gen. William Wray, March 17, 1980; Interview with Alan Van Den Berg, Dec. 6, 1979.

15. Acting Deputy Chief of Research and Development Maj. Gen. John Paley, to the Chief of Research and Development, Jan. 19, 1957.

16. Secretary of the Army Wilbur Brucker, Letter to President Dwight Eisenhower, Feb. 8, 1957.

17. Interview with Melvin A. Rosen, Jan. 29, 1963; *Chief of Engineers Summary of Major Events and Problems*, FY 1957, *New York Times*, April 30, 1957, pp. 13–14.

CHAPTER 4

# FORT GREELY

Long before the SM–1 began its career as a training and research facility, the Army Reactor Branch had begun work on plans for the first field nuclear power plant. From the start, Alaska remained the leading candidate for the first remote plant. By early 1955, the Reactors Branch had selected Fort Greely, situated 100 miles southeast of Fairbanks, as the preferred site. The Branch justified its decision on the grounds that it would "satisfy military requirements for electricity and heat with logistic efforts greatly reduced relative to existing conventional systems." The plant would also "serve as a transitional test facility to gain operational experience under conditions of extreme temperature (+90 degrees to −60 degrees) in order to evaluate the nuclear power concept under arctic field conditions." The data gained would "complement rather than duplicate that gained from the Fort Belvoir plant (APPR–1)."[1]

The Army anticipated that preliminary design and site work would start in FY 1956 and construction at Fort Greely would begin during FY 1957, at an estimated cost of $4,127,000. In responding to the proposed facility, the Bureau of the Budget had questioned the plan on the grounds that no Army base in Alaska was sufficiently remote or required replacement power facilities that would justify a nuclear plant. The Bureau also pointed out that the project had not been coordinated with the AEC on the commission level. In its answer, the Army maintained that a nuclear plant "will markedly relieve logistic effort in supplying heat and power at a strategically important installation." It also cited experience to be gained "under arctic conditions necessary to complete evaluation of nuclear power plant concept for arctic military use." Moreover, the Army said that the Reactor Branch had informed the AEC of its plans through internal consultation and would make formal notification through the Military Liaison Committee.[2]

Despite its initial planning and the momentum gained from the start of construction on the SM–1, the Corps of Engineers did not receive official approval of the Alaskan nuclear plant from the Secretary of Defense until January 1956. The Pentagon then included the facility in its FY 1957 Military Public Works Program, which it submitted to Congress on January 15. The plans called for a facility to provide heat and electricity for a military base utilizing a reactor modeled after the SM–1.[3]

With the procurement bill beginning to make its way through Congress, the Military Liaison Committee advised the AEC of the Pentagon's decision to build the initial field facility. It also indicated that the Army Reactors Branch would seek advice and assistance from the Commission at the appropriate time in accordance with agreed-upon policies. In informing Alaskan Delegate E. L. Bartlett about the Defense Department's decision in April 1956, the Army explained that the Reactor Branch had not yet determined the exact location of the proposed plant. However, it said the Branch was making the selection "with a view to providing the maximum benefit to the military forces from viewpoints of operational experience and potential logistics savings." According to the Army, the Corps of Engineers would contract the project "with a suitably selected industrial firm or group of firms" and would finance the construction "wholly from funds appropriated to the Department of the Army."[4]

To make a final determination of the location for the nuclear power plant and to assist the Alaska District Engineer on negotiations for its architect design contract, Col. Lambert had sent Lt. Col. Gribble to Alaska in March 1956. Gribble later recalled that the Reactor Branch had the confidence to begin work on the first field unit while the SM–1 was still under construction because of the "careful development" of the prototype plant. The planning of the Alaskan facility was designed to take advantage of the experiences gained as work on the SM–1 progressed and then went critical. Consequently, although the plant in Alaska would be similar in design and size to the Belvoir reactor, the Reactor Branch hoped "to extrapolate from that state of art by doubling the power removed from the system" so that the capacity would be twice the prototype plant with only minor changes in the basic design.[5]

A May 9, 1956, Staff Study by the Reactor Branch explained that the doubling of the capacity and the fact that the facility would be the first nuclear reactor "designed for and operated under arctic conditions" signified that the Alaskan plant would "not be a standard, production-type item." As a result, the Study proposed that "the plans for design, construction, and operation of this plant should proceed on the basis that the plant is a transitional test facility, until such time as the plant has been completely proven." The Staff Study then looked at two approaches to procurring the plant. The first one resembled the process by which the Army was building the SM–1. The District Engineer in Alaska would negotiate a contract for design and construction of the conventional aspects of the plant, while the

AEC, using Army money, would negotiate for design, construction, and delivery to the site of the reactor system. As an alternative, the District Engineer could negotiate "a single contract for design, construction and test operation of the entire plant" with a firm that had demonstrated the capacity to produce a reactor system, supervise the overall plant design, prepare and secure approval of the Hazards Study for the AEC, and supervise the building and testing of the facility.[6]

The Study pointed out that the alternative proposal eliminated the disadvantages of split responsibilities inherent in the construction of the SM–1. Positively, the new approach would develop a contractor with the ability to produce a complete plant, who would have developed a working relationship with the Corps of Engineers, and who "would be responsive to sudden demands to take on additional work." The Reactor Branch considered this important in light of its belief "that a crash program to build reactor plants is inevitable, after feasibility and reliability are proven by [SM–1] operation." To the Staff, the new approach had only three disadvantages: (a) it was "an unusual, though legal procedure" for the Engineers; (b) a change in procedure "would have an unsettling effect"; and (c) it would require closer field inspection by the District because it "could not rely on the Prime Contractor to inspect his own work or work done by his sub-contractor." On balance, the Staff Study, which Lambert signed, concluded that the Alaskan plant should be procured under the alternate approach.[7]

A proposed personnel training program for military nuclear power plants accompanied the Staff Study. The program, as described, dealt with both the overall support training program that the Reactors Branch and the Nuclear Power Branch were then developing at Fort Belvoir and plans for the operation and maintenance of the Alaskan plant, now officially planned for Fort Greely. For the Alaskan facility, a program would be developed "through sound integration of experience with the nuclear power plant here at Fort Belvoir and experience with conventional plants in Alaska." The Memo pointed out that the SM–1 had as one of its main purposes the instruction of operators and maintenance personnel. It was expected that the plant's start-up crew would include military personnel who would then become "a cadre of instructors and plant supervisors for such field units as the Fort Greely plant."[8]

According to the training program proposal, the balance of the crew for the Greely plant would be recruited during the summer of 1957. They would receive their training at the beginning of 1958 and then be sent to Alaska in April with the rest of the personnel to participate in the final installation, start-up, and test operation of the facility. Following the six-month trial, routine operation was expected to commence in the fall of 1958.[9]

While planning progressed for the Greely facility, the Reactor Branch continued its interest in other military utilizations for small nuclear reactors.

Out of its early inquiries to the services, the Engineers elicited an interest from the Transportation Corps for a nuclear power overland cargo hauler designed to provide logistical support in desert or arctic wastelands. In August 1955, graduate students at the Oak Ridge School of Reactor Technology had produced a design for a nuclear powered train. It was to consist of up to 10 cars each with a 15-to 30- ton capacity, rolling on wheels with tires 10′ in diameter with a tread 4′ wide. Diesel-powered versions then under consideration had to sacrifice considerable payload in order to carry the necessary quantity of fuel, but a nuclear-powered version would extend the limit and cargo capacity of the train.

Following up on the Oak Ridge school's outline, in December 1955, the Transportation Corps had initiated a contract with the Nuclear Development Corporation (later, the United Nuclear Corporation) to study the application of nuclear power to their rolling stock. In a briefing to the Assistant Secretary of the Army for logistics on February 15, 1956, the Transportation Corps indicated that the $100,000 design and feasibility study would be completed in six to eight months. At the same time, the Transportation Corps explained that the AEC was continuing its own studies and had agreed to fund the reactor and generator for the snow train. Army Operations Research Office figures indicated that operating costs for a nuclear vehicle would be very low "as compared to present arctic transportation costs."[10]

In the same meeting, the Transportation Corps also discussed its investigation of the possible use of nuclear power for locomotives, large trucks, lighters, and landing barges. The briefing officer pointed out the problems of building a nuclear-powered tank. These included the tank's excessive weight (at least 50 to 60 tons), the difficulty of incorporating the reactor and its necessary shielding behind adequate armor in an acceptably small package, and the potential for radioactive contamination of a battlefield "if enemy anti-tank fire blew up the reactor or caused a loss of reactor control which would result in a low order atomic explosion." Nevertheless, he concluded that the military's requirement for small reactors would be followed by reduced reactor weight and sizes. Consequently, the Army was studying the possible applications of such small, light reactors to exploit them when they were developed.[11]

By mid-September, the Army informed the AEC's Division of Reactor Development that the initial phases of the contract with the Nuclear Development Corporation had been completed, including the determination of the technical feasibility of the conceptual design of a reactor for the land train. According to the Army, the study indicated that it was technically feasible to develop a reactor system for the train. The Army advised the Reactor Development Division that its Operations Research Office was now examining the economic and military advantages of applying this technology to mobile military equipment and the Office expected to present its findings to the Defense Department by the end of November.[12]

The Army indicated that if the current studies concluded development was desirable, it expected to make a formal request to the AEC for prototype development by December. In this light, the Army was alerting the AEC "to the possibility that a firm requirement" for such a program would be made in order to allow the Commission to plan for the item in its FY 1958 budget. According to consultation with an industrial organization that did nuclear research and development, approximately $1.75 million would be needed "to establish the program on a firm basis and minimize overall research and development time."[13]

Most of the Army's attention during 1956, however, focused on larger, fixed power-generating facilities to be operated in overseas locations. In January, Lambert assigned Gribble to explore the technical feasibility of building a nuclear power plant to meet the increasing power demands of U.S. forces stationed in Okinawa. Although the Reactors Branch ultimately did not include the site on its roster of power plants, the Army did get drawn into the considerations for building such a facility under the auspices of the Defense Department. On July 9, the Army's Research and Development Office received a request from the Defense Department's Electrical Equipment and Systems Branch to help formulate the Pentagon's position on the proposed plant. The invitation advised the Office of the complexity of the subject, including the technical and economic problems, the matter of future ownership of the facility, and the type of reactor to be built. According to the letter, "the most important of all factors will be national policy" rather than engineering or economic considerations.[14]

The proposal to build a nuclear power plant on Okinawa had resulted from the request of the Far East Command for additional power facilities to meet anticipated future demands for more electricity. In considering the request, the Chief of Civil Affairs and Military Government had concluded that a facility combining nuclear power and conventional power would cost $35.7 million and would produce 42 megawatts of electricity. The nuclear portion would produce 20 megawatts at a cost of $27.4 million, whereas the conventional plant would produce 22 megawatts at a cost of $8.3 million. In contrast, a completely conventional plant would cost $13.8 million. The AEC had agreed to provide $20 million in research and development funds and had already let a contract for preliminary design work. The Office of Government Assistance and Relief in Occupied Areas was to provide $13.8 million of the funds because the Army's Deputy Chief of Staff for Logistics was not willing to commit Military Construction Appropriation funds to the project.[15]

Advocates of the power project had several reasons for wanting to include a nuclear facility in the new power-generating complex. They wanted to ensure availability of uninterrupted power in the event that a war emergency cut supply lines, to reduce logistical requirements for overseas supply of fuel during a war, to demonstrate the United States' use of atomic power for peaceful purposes, to help the Okinawans improve their economy, and to test the operation of nuclear power reactors in tropical climates. (In fact, Okinawa did not offer a true tropical climate.) In the meeting of July 2, the Army's Chief of Civil Affairs and Military Government also expressed his belief that an important aspect of the nuclear power plant would be to stimulate friendly feelings toward U.S. citizens. Although representatives of the other services did support the requirement of additional power-generating capacity on Okinawa, none expressed a preference for one type of facility over the other.[16]

In regard to control of Okinawa as it related to operation of the nuclear power plant, the Navy representative pointed out that the United States had stated it did not intend to remain on Okinawa and indicated he thought Japan would object to the facility. The representative of the Chief of Civil Affairs and Military Government expressed his belief that the transfer of title for the plant to the Ryukuan Government was of little importance because he thought the United States would still be controlling Okinawa when the plant wore out in an estimated 30 years. On his part, Col. Lambert indicated the nuclear plant would be modeled after the commercial plant that the Westinghouse Corporation was building at Shippingport, Pennsylvania. He estimated that the plant would take a year to design and two years to build and he believed that legal problems involving transfer of fissionable material did not constitute an insurmountable problem. At the same time, he did not assign a priority to the facility in the Army Reactor Branch's development and construction program. Consequently, although they joined the group created to consider the project, the Engineers did not encourage the building of the plant at that time.[17]

That summer, however, the Army began formulating an official position in regard to the further establishment of fixed nuclear power plants at overseas bases and the use of small mobile plants on a less permanent basis. On August 7, through the Army's Research and Development Office, the Military Liaison Committee advised the Reactor Branch that the military's primary interest in nuclear power facilities focused on bases located in U.S. territories such as Alaska, Hawaii, and Puerto Rico "with possible extension of the program to bases on foreign territory" occupied by U.S. forces. The Research and Development Office indicated in its Memo that the requirements for overseas plants could "be established either on the basis of economy or on the basis of military advantage."[18]

In considering the economics of nuclear power, the Office pointed out that significant savings in fuel costs had to be realized to justify nuclear plants because they required a much higher initial investment than conventional power plants. The Office, therefore, suggested that nuclear power plants could generally be justified on the basis of economy "where conventional fuel cannot be delivered by surface means." Military advantages included "the reduced logistics load for transportation systems, the resistance of the nuclear power plant to damage because of its inherent heavy construction, and the suitability of the nuclear power plant for underground construction because of its lack of combustion gases." The only disadvantage advanced in the Memo cited "the possible contamination of a small area in the case of destruction of a nuclear power plant."[19]

The Research and Development Office suggested that most desirable locations for nuclear power plants in terms of economics were arctic sites because they "are inaccessible or accessible with great difficulty for surface transportation, and conventional fuel is a major logistical problem." At the same time, the Office pointed out that nuclear power plants for larger, less remote installations in the arctic would be competitive with conventional power plants "while not completely justifiable economically." On the other hand, in a criticism that would complicate the final selection of Fort Greely as the site for the first nuclear field facility, the Research and Development Office advised that reactors "located near sources of fuel or easily accessible by tankers or railroads cannot be justified on an economic basis at this time."[20]

For military purposes, the Research and Development Office indicated that the most desirable locations for nuclear power plants would be "at the ends of long and exposed supply lines, particularly in areas suitable for large scale logistic buildup." Remote sites could also be considered in assigning military advantages for a nuclear plant when "the only supply source is airlift." Reflecting the Defense Department's justification for the proposed Okinawa facility, the Office suggested a military advantage for nuclear plants "to support defense of overseas United States possessions in the event such possessions are cut off from normal logistical support."[21]

In response to the Memo from the Office of Research and Development, on August 17, the Army Office of Operations proposed that the Army actively support reactor programs for small nuclear power plants in remote locations "where airlift is the primary means of logistical support"; for example, radar installations in the arctic. In regard to less isolated locations such as the proposed plant at Fort Greely, the Army justified the facility not on economic grounds but because it was "suitable for training and development purposes." The Operations Office also proposed that the Army consider other factors in siting nuclear power plants. These included areas in which the Army had long-range occupation plans, areas that had strategic importance, the impact the cost of the nuclear power plant would have on other Army programs, and the degree to which local fuel supplies would be available to support a conventional power plant in the event the area became isolated in an emergency situation.[22]

Out of this exchange of correspondence and other memoranda, the Army formulated and then issued on August 21, a "Statement of Army Requirements for Nuclear Reactors," which the Army Research and Development Office then sent to the Assistant Secretary of Defense for Properties and Installations. In the statement, the Army included criteria for evaluating nuclear power requirements and general planning guides for utilization of nuclear power plants. The criteria for evaluating requirements for the facilities were subdivided into economic and strategic factors. Economic criteria in general order of importance included the cost of delivered conventional fuel, installed plant capacity required, load characteristics that determine plant factor, and plant life or planned period of operation. Strategic criteria included the strategic importance of the proposed installations during wartime, dependence of installation on vulnerable supply lines for resupply of conventional fuel, the planned expansion of the installation from peacetime strength that would be required to carry out its wartime mission, minimum power capacity required for the operation in emergency situations, the period of operation anticipated for the installation, and the degree of U.S. control over the area in which the operation existed.[23]

The Statement proposed building two types of plants. In the U.S.-controlled territories or areas the United States expected to control for at least 10 years, fixed-type nuclear facilities should be constructed "to carry the base load of the installation at its currently planned peak expansion level." In transient locales (5 years or less), the Army proposed to build mobile power plants of several capacities "to permit flexible use under varying conditions." Likewise, in areas not under direct U.S. control, such as Korea, North Africa, and NATO countries, mobile plants should be developed and deployed. The Statement then looked at power requirements for Army overseas installations that could be met with nuclear power plants of one type or the other, subdivided into three categories of priorities. Fort Greely was the only location listed in the first category.[24]

By the end of October, however, the Army Reactors Branch had still not been able to make a final determination of Fort Greely as the site for its first field reactor even though the Corps of Engineers had expressed its preference for that locale. Among other factors influencing their decision, the Army had an Artic Test Center on the base and conducted all types of cold weather tests and experiments there. In any event, the suitability of constructing the reactor at Greely was discussed at length in a meeting on October 30 in the office of the Assistant Secretary of Defense (Supply and Logistics). This meeting was called to consider the $3 million budget request for FY 1958 to begin construction of the Alaska facility.[25]

In a review of the budget, the Office of Assistant Secretary of Defense (Research and Development) had raised questions about the advisability of building both the Army's proposed reactor at Greely and an Air Force reactor the next year at Sparrevohn. In justifying the proposal for its plant at the October 30 meeting, Air Force representatives explained that the only means of supplying the base was through an air lift that delivered approximately 11,000 barrels of fuel a year. This constituted about 75 percent of the base's total logistic requirements. Because of the short runway at the base, the airlift was difficult and a C-124 cargo plane had recently crashed. In this light, the Air Force hoped to reduce air traffic by 75 percent, if at all possible. According to the service's representatives, a nuclear power plant at the site would greatly reduce the exposure of airplanes to mishaps. At the same time, the Air Force indicated that the Army should begin work on its plant first because of better accessibility at Greely. Once their own appropriations were received, however, the Air Force said design work could begin quickly with actual construction expected to start in the spring of 1958 and completion of the facility expected the following summer.[26]

On his part, Col. Lampert acknowledged that the Fort Greely site could not be completely justified economically because a fuel pipeline now served the base. Nevertheless, he pointed out that it was highly desirable to build the first arctic reactor at an accessible location so that operation and technical problems could be worked out prior to installation of a reactor at a site accessible only by air—the situation at the Air Force's proposed facility. He also indicated that the Defense Department would make a final decision on the site by January or February 1957.[27]

In the meantime, he informed the meeting that the Army had entered into a design contract for the Alaska reactor and explained that the size and type finally selected would represent the maximum reasonable extrapolation of the SM–1. In response to a question about other types of nuclear plants under consideration, Lambert said the Reactor Branch would work next on the development of a gas-cooled reactor that would probably be suitable for smaller plants, perhaps even one that would fit onto a 30-ton trailer. According to Lambert, the first gas-cooled reactor would be an experimental facility, not a prototype.[28]

Despite the meeting, the fate of the Greely plant remained in doubt. Only a week later, on November 6, the Army's Research and Development Office learned that the Defense Department's Coordinating Committee on Atomic Energy, which was reviewing the budget request for the Greely facility, might recommend that the project be dropped in favor of building the proposed Air Force power plant. The Research and Development Office, in turn, was prepared to "protest strongly" to the Coordinating Committee if it did issue an unfavorable report on the Greely project.[29]

The proposed reactor had another opponent; this one within the Army Command in Alaska. The Army Commander there refused to include the proposed nuclear power plant in the budget he submitted to Congress for capital improvements because he felt its cost would affect other projects that he favored more. In a letter transmitted through the Commanding General, Alaskan Command, he stated his position to the Chief of Engineers, Lt. Gen. Samuel Sturgis. In contrast, the Commanding General, as an Air

Force officer, perceived that nuclear power might answer the power needs of the Distant Early Warning (DEW) Line radar stations and argued on behalf of the Greely facility. In the end, Gen. Sturgis solved the problem by including the Greely plant in the Corps of Engineers' budget.[30]

In the meantime, the Reactor Branch had negotiated a design contract with ALCO Products in July 1956 for the design of the Greely facility (SM–1A). By June 1957, when the Defense Department had approved the Greely facility as part of the FY 1957 military construction program, ALCO had completed the detailed design of the plant and the Army had submitted a request to Congress to boost the authorization for the reactor from $4.1 to $6.7 million. And, in January 1958, ALCO completed the design of the plant, modeled on the SM–1 at Belvoir, but intended to supply heat as well as electricity to the base.[31]

By then, the Reactor Branch had undergone a major change in leadership. In the summer of 1956, Col. Gribble had been reassigned from his position as Col. Lambert's assistant to Assistant Alaskan District Engineer. He was thus able to assist the Alaskan District Engineer in the review of the architect engineer's submissions and plans for the Greely plant as well as help prepare the project for bidding. In fact, in this position until 1958 and then as the District Engineer for Alaska from 1958 to 1960, Col. Gribble functioned as de facto Project Engineer for the SM–1A.[32]

After managing the reactor program for five years, Lampert also left the Army Reactors Branch in August 1957 upon receiving an assignment to the National War College for the 1957–1958 academic year. As his replacement, the Army selected Col. Donald Williams, a graduate of the U.S. Military Academy and Iowa State University with a master's degree in Civil Engineering. His Army service in the atomic energy field included duty with the Manhattan District, the Special Weapons Group at Sandia Base, New Mexico, and the Military Liaison Committee to the AEC, where he was the military deputy to the Assistant to the Secretary of Defense for Atomic Energy. Before entering the Industrial College of the Armed Forces in August 1956, he spent six months with the Army Reactors Branch.[33]

In helping negotiate the contracts for the SM–1A, Col. Gribble followed a different procedure from the one used to build the SM–1 at Fort Belvoir. With the prototype plant, the Engineers had entered into a package deal with ALCO, which assumed responsibility for both the design and construction of the facility. With the Fort Greely plant, the Engineers first let a contract with ALCO for the design of the facility and then the Alaska District advertised bids for its construction from a list of contractors experienced in working under the adverse Alaskan environment. Peter Kiewit Sons' Construction Company of Seattle underbid eight other firms and on April 26, 1958, received a contract to build the plant at a cost of $4,897,217. In turn, Kiewit subcontracted the nuclear system within the plant to ALCO Products thereby reversing ALCO's role in building the Belvoir facility, where it had served as the prime contractor and had sublet the nonnuclear system. In establishing a typical contractor/subcontractor relationship, Kiewit, for the first time, built a nuclear power plant under normal operating procedures of the Corps of Engineers.[34]

The design for the SM–1A called for the plant to generate 4,000 kilowatts, half for the electrical needs of the base and half in the form of steam for heating purposes. As with the Fort Belvoir plant, the reactor core was to contain thirty-eight stationary fuel elements and 7 control rods. The plant was expected to supply all but peak loads for Fort Greely's physical plant and 2,000 people. The additional power the base would require was to be supplied by an adjacent diesel power plant that would operate in parallel to the nuclear plant. In turn, the nuclear-diesel complex would be in parallel with the Golden Valley Electrical Association, a combination of civilian and military plants operating in the Fort Greely area.[35]

Like the SM–1, with its proximity to Washington, the SM–1A's vapor container was an essential part of the Fort Greely plant due to its location in an area of several hundred military. In contrast to the SM–1's dome-shaped vapor container, the one at Fort Greely had a cylindrical-shaped exterior steel shell that enclosed a cylindrical concrete tank. The 2 feet 10 inches between the outside of the concrete tank and the inside of the steel tank was filled with water. In addition to providing shielding, the water served as a heat sink to cool vapors that would fill the concrete tank in the event of accidental primary systems rupture.

A second unique feature of the SM–1A was the manner of obtaining the condenser cooling water used to condense the turbine exhaust steam. Whereas the SM–1 was able to obtain its cooling water from the nearby Potomac, Fort Greely had no convenient surface water supply. As a result, the plant had to rely on water constantly recirculating through deep wells to condense the steam. But, because of the hardness of the well water, the system required the addition of a water-softening process, which represented another step in the development of nuclear power plants.

Despite being based on an already working design provided by an experienced company, the SM–1A did not have an easy road to completion. The construction project's "stormy history," as Gribble described it, had nothing to do with the "nucleonics" of the plant. Rather, the problems were Alaska's short construction season, which ran only from May to September, the long supply lines from the lower forty-eight states, and, most important, a major labor stoppage the second year of construction. Because the construction contract was awarded late, Kiewit could not begin work until June 4, 1958. During the first year, the company was able to mobilize its resources, place orders for long lead-time material and equipment, and dig the foundation hole so that it would be waiting for the start of the 1959 building season. By the time work ended the first fall, approximately 10 percent of the on-site construction had been accomplished.[36]

Construction resumed according to schedule the next spring and, by the end of June, the contractor had moved into the hole with his concrete, reinforcing bars, and heavy bundles. However, with about 35 percent of the work on the plant completed, all organized labor involving construction crafts and trades went out on an Alaska-wide strike. A settlement was reached only in mid-September, when the construction season had just about come to an end. As a result, Kiewit had slipped almost a whole year in its construction schedule.[37]

Since the strike was due to circumstances beyond the contractor's control, Col. Gribble negotiated an agreement with Kiewit for a one-year extension. In addition to the strike, however, time-consuming procurement and delivery of materials to the site contributed to the slow completion of the SM–1A. Materials had to be trucked to Fort Greely either on the Alaska Highway from Seattle or from the port of Valdes or airlifted to the Big Delta Airfield near the base. Shipment of the larger reactor components, including the pressurizer and reactor vessel, from New York involved cross-country movement by rail to Seattle and then by ship to Valdes and finally by truck to Fort Greely. Some of the more delicate parts, such as the core support structure and the radiation monitoring board, did not withstand rough handling and had to be shipped back to the suppliers for repair and reshipped in sturdier crating. In particular, the damage to the core support structures was very serious and caused a lag in the construction schedule.[38]

Another problem had to do with the procurement of certain stainless steel fittings and related equipment peculiar to a nuclear power plant unavailable in Alaska. As a result, it required a carefully arranged shipping schedule to prevent delays in construction. In addition, the requirement of 30,000 gallons of distilled water for system cleaning, testing, and operation necessitated the construction of a temporary still. Adding to the slowness of construction was the need to mix over 2,500 cubic yards of concrete at Fort Greely and deliver it to the site in trucks.[39]

Due to these delays, the power plant was only about 87 percent to 90 percent complete by the time Col. Gribble left Alaska in July 1960. Nevertheless, in anticipation of completing the project during that summer, the Reactor Branch had dispatched part of the operating crew from Fort Belvoir in April to prepare for the plant's performance tests. In June, the first nuclear fuel arrived in 11 packages. Nevertheless, delays continued to plague the final stages of construction and the plant's nonnuclear testing was not completed until early October.[40]

Most of the problems centered around faults in the nuclear instrumentation, thus the beginning of power tests was pushed back several times. With the completion of the nonnuclear testing, it had been expected that loading of the core would take place in October. Because of the malfunctions with the instrumentation equipment, however, ALCO returned it to the company's critical facility for preliminary testing and checkout. It then returned the equipment to Fort Greely on October 30. When subsequent performance tests revealed further deficiencies, General Electric sent three men to Fort Greely. Arriving on December 1, they worked with the contractor and the operating crew to bring the instrumentation up to specification standards.[41]

Expectations of a quick solution to the problems were thwarted when trouble in the circuit boards for the nuclear instrumentation forced ALCO to return them to the General Electric plant in San Jose, California. When they could not be repaired, the company had to build new circuit boards, which were scheduled for delivery in late March. This forced ALCO to push back core loading to April 15, 1962. In a reversal of form, however, the equipment arrived earlier than expected; all systems checked out and the SM–1A went critical on March 13, 1962.[42]

From their arrival in June 1960, until the SM–1A finally did go critical, the crew from Fort Belvoir worked with the contractor to eliminate the flaws in instrumentation and prepare the plant for operation. Using their knowledge gained from working on the SM–1, the crew—one officer, one warrant officer, and seventeen enlisted men—worked in four shifts once the reactor became operational. Eight men operated the plant, five attended to maintenance matters, two to health, physics, and chemistry laboratories, and two to administrative matters.

Following completion of core loading, the reactor pressure head was closed on March 16, and by March 19 the first three low-power nuclear tests had been satisfactorily completed. The initial series of low-power and noncritical tests were completed on March 24. On April 23, the SM–1A assumed Fort Greely's entire electrical and steam load. In addition, the reactor put a small 50-kilowatt load of electricity into the commercial grid supplying power to Delta Junction. The next morning, the power plant went to full power for the first time. In reaching its rated load of 2,000 kilowatts, minor problems with the steam turbine were corrected and heat balance and nuclear instrument calibration tests were performed.[43]

Following the attainment of full power, the SM–1A began a month of full-power shakedown testing. Although the reactor functioned without problem, the nuclear instrumentation continued to cause trouble; General Electric finally supplied new circuit boards to replace the original boards that had undergone modification and adjustment. The panels were installed during the first four days of June and the Alaska Engineer District, with the assistance of the Nuclear Power Field Office from Fort Belvoir, performed a Pre-Final Inspection of the power plant on June 5. The Office of the Chief of Engineers held a three-day inspection and on June 15, 1962 it assumed operational responsibility of the SM–1A.[44]

The plant's problems had not come to an end, however. After a scheduled shutdown for regular maintenance, the crew began start-up operations on June 26, only to discover that the high-pressure seals in the Mechanical Seal

Pump were discharging large amounts of water. The Nuclear Power Field Office contacted the manufacturer of the equipment and a technical representative from the company arrived on June 29. When the pump was removed from the primary system, the crew found that the primary seal had been badly damaged. Repairs were completed on July 3 and the crew put the plant back on-line July 4.[45]

Although the seal continued to cause additional problems in the coming months, the reactor itself became the source of a malfunction on August 13 when two control rods got stuck while 4 to 5 inches out of the core. The crew made several adjustments and the reactor went back on-line September 3. Following this problem, on September 13 the plant's Hot Waste Discharge System was put into operation and for the first time, liquid hot waste was discharged into Jarvis Creek about a mile from the plant.[46]

In operation, the plant stored radioactive liquid wastes underground in three tanks until radioactivity had sufficiently decayed to permit the safe discharge to an off-site waste-dilution station. It was there mixed with well water before being discharged into the creek. During regular operation, at the start of each summer's discharging operation, water samples taken upstream of the discharge point showed more radioactivity than those from below the discharge point. The apparent paradox occurred because melting snows that had accumulated natural radioactivity during the winter fed the creek while the SM–1A operators used well water to dilute the waste discharge from the plant. In some sense, it might be argued that the waste from the SM–1A was diluting the stream's natural radioactivity.

In any case, by November 20, 1962, the SM–1A had been on-line a total of 1,658 hours out of a possible 1,920 hours, beginning on September 3, making it "on-line" 80.6 percent of the time. Nevertheless, minor problems continued to plague the SM–1A. One unscheduled shutdown occurred on September 24, when a short developed in the electrical circuits leading to the control rod assembly. The problem was caused by water leaking into the rod drive pit from a leaking gasket in the steam generation inspection port; repairs required a day. A second problem with the control rod mechanisms caused a shutdown in October. However, the operating crew gradually worked out the bugs in the plant and during the last quarter of 1962 the plant was on-line 96.3 percent of the time.[47]

After a year of routine operation and repairs, planning was initiated to install the second core during the spring of 1964. Initially scheduled to take place in February, the core was not installed until April and the plant went back on-line on April 21. Before that took place, however, a major earthquake on March 27 caused heavy damage to southern Alaska but did only minor damage around Fort Greely. The plant's vapor container did sway and the sloshing in the various tanks set off both high and low water alarms. Fortunately, the plant itself suffered no damage.

On July 1, 1964, the Corps of Engineers transferred operational control of the SM–1A to the U.S. Army, Alaska. In preparation for the transfer, the Nuclear Power Field Office conducted an on-site safety review covering all aspects of plant operation and maintenance. As a result of the transfer of control, the SM–1A was a fully operational military reactor when, on August 9, 1964, it set a record for an extended power run by a military nuclear power plant of 2,630 hours. This surpassed the record of 2,502 continuous hours set in February 1963 by the PM–2A at Camp Century, Greenland and the record of 2,630 hours set in April 1964 by the PM–1 at Sundance Air Force Station in Wyoming. When the reactor was finally shut down for maintenance on August 14, it had been on-line continuously for 2,750 hours and 42 minutes since the installation of its second fuel core on April 21. By comparison, the Yankee Atomic Power Plant located at Rowe, Massachusetts, had been operating continuously for 3,480 hours, the U.S. record. Nevertheless, the SM–1A's record demonstrated that, despite initial problems, the plant could provide reliable power for a remote military base.[48]

## NOTES

1. *Summary of Army Project for Nuclear Power Plant in Alaska,* n.d. [Jan. 1955].
2. Ibid.
3. Army Research and Development Office, to Chairman, Military Liaison Committee, Jan. 23, 1956.
4. Secretary of the Army, to Delegate E. L. Bartlett, April 6, 1956.
5. Interview with Brig. Gen. William C. Gribble, Jan. 23, 1963.
6. Col. James Lambert, *Staff Study,* May 9, 1956 (Document).
7. Ibid.
8. *Personnel Training Program for Military Nuclear Plants,* n.d. [May 9, 1956], Box 6.
9. Ibid.
10. Letter from Army Director of Special Weapons to W. Kenneth Davis, Director, Division of Reactor Development, Atomic Energy Commission, September 14, 1956, Box 5; Department of the Army, *Memorandum for Record,* Feb. 15, 1956.
11. *Memorandum for Record,* Feb. 15, 1956.
12. Army Director of Special Weapons, to Division of Reactor Development, Atomic Energy Commission, Sept. 14, 1956, Box 5.
13. Ibid.
14. Chief, Electrical Equipment and Systems Branch, to Army Office of Research and Development, n.d. [July 9, 1956]; Gribble interview.
15. Chief, Atomic Division, Army Office of Research and Development, Col. Robert E. Coffin, *Memorandum for Record,* July 16, 1956.
16. Ibid.
17. Ibid.
18. Army Office of Research and Development, Memo to the Deputy Chief of Staff for Operations, Aug. 7, 1956.

19. Ibid.
20. Ibid.
21. Ibid.
22. Army Director of Operations, Response to Office of Research and Development, Aug. 17, 1956.
23. Army Office of Research and Development, "Statement of Army Requirements for Nuclear Power Reactors," Aug. 21, 1956.
24. Ibid.
25. Chief, Electrical Equipment and Systems Branch, Department of Defense, to the Army Research and Development Office, Nov. 5, 1956; Army Office of Research and Development, *Memorandum for Record,* Nov. 13, 1956.
26. Ibid.
27. Ibid.
28. Ibid.
29. Army Office of Research and Development, *Memorandum for Record.*
30. Gribble interview; Maj. Gen. David H. Tulley, to K. J. Deacon, July 31, 1974.
31. Gribble interview.
32. Ibid.
33. Secretary of the Army, to Lewis Strauss, July 10, 1957.
34. Gribble interview.
35. The discussion of the building and operation of the SM–1A came from the documentary and printed material in the SM–1A files in Box 6 of the ANPP records, unless otherwise cited. In particular, see the following: Francis Wolak, "Army Field Nuclear Power Plant," *The Military Engineer* (November-December 1961): 434; and U.S. Army Engineer Reactors Group, "The Army Nuclear Power Program, Five on the Line," July 1962, Box 3 (hereafter cited as "Five on the Line.").
36. Gribble interview; Francis Wolak, "Army Field Nuclear Power Plant," *The Military Engineer* (Nov.–Dec. 1961).
37. Wolak, "Army Field."
38. Ibid.
39. Ibid.
40. Gribble interview; Army Corps of Engineers, *News Bulletin #1,* Sept. 7, 1961, cited hereafter as *News Bulletin.*
41. Ibid.; *News Bulletin #2,* Nov. 17, 1961; *News Bulletin #3,* Dec. 15, 1961.
42. *News Bulletin #5,* Feb. 15, 1962; *News Bulletin #6,* March 15, 1962.
43. *News Bulletin #7*, April 13, 1962; *News Bulletin #8,* June 1, 1962.
44. *News Bulletin #9,* July 2, 1962.
45. *News Bulletin #10,* Aug. 7, 1962.
46. *News Bulletin #12*, Nov. 15, 1962.
47. *News Bulletin #12,* Nov. 15, 1962; *News Bulletin #13,* Dec. 15, 1962; *News Bulletin #14,* Feb. 5, 1963.
48. *News Bulletin #26*, Aug. 14, 1964.

1. SM–1 under construction at Fort Belvoir, Virginia. Welders are putting finishing touches on exterior welds of the vapor container ring segments. Each seam in the the 78″ steel shell was welded on both sides and the welds were then x-rayed to assure uniform bond.

2. SM–1 at Fort Belvoir, Virginia.

3. Overhead view of core structure in SM–1 as technicians make final installation. Note 7 × 7 grid, with corners missing, that provides space for fuel elements and control rods.

4. Control room of SM–1, Fort Belvoir, Virginia.

5. Components of the Portable Nuclear Power Plant PM–2A sitting along Ramp Road, Camp Tuto, in preparation to move to Camp Century, July 15, 1960.

6. The vapor condenser of the PM–2A Nuclear Power Plant being moved by Army Engineers into the power trench at Camp Century, Greenland.

7. SM–1A at Fort Greeley, Alaska.

8. PM–3A at McMurdo Sound, Antarctica.

9. PM–1 at Sundance, Wyoming.

10. Nuclear Power Barge SURGIS in the Panama Canal Zone.

11. The SL–1 Boiling Water Reactor in Idaho.

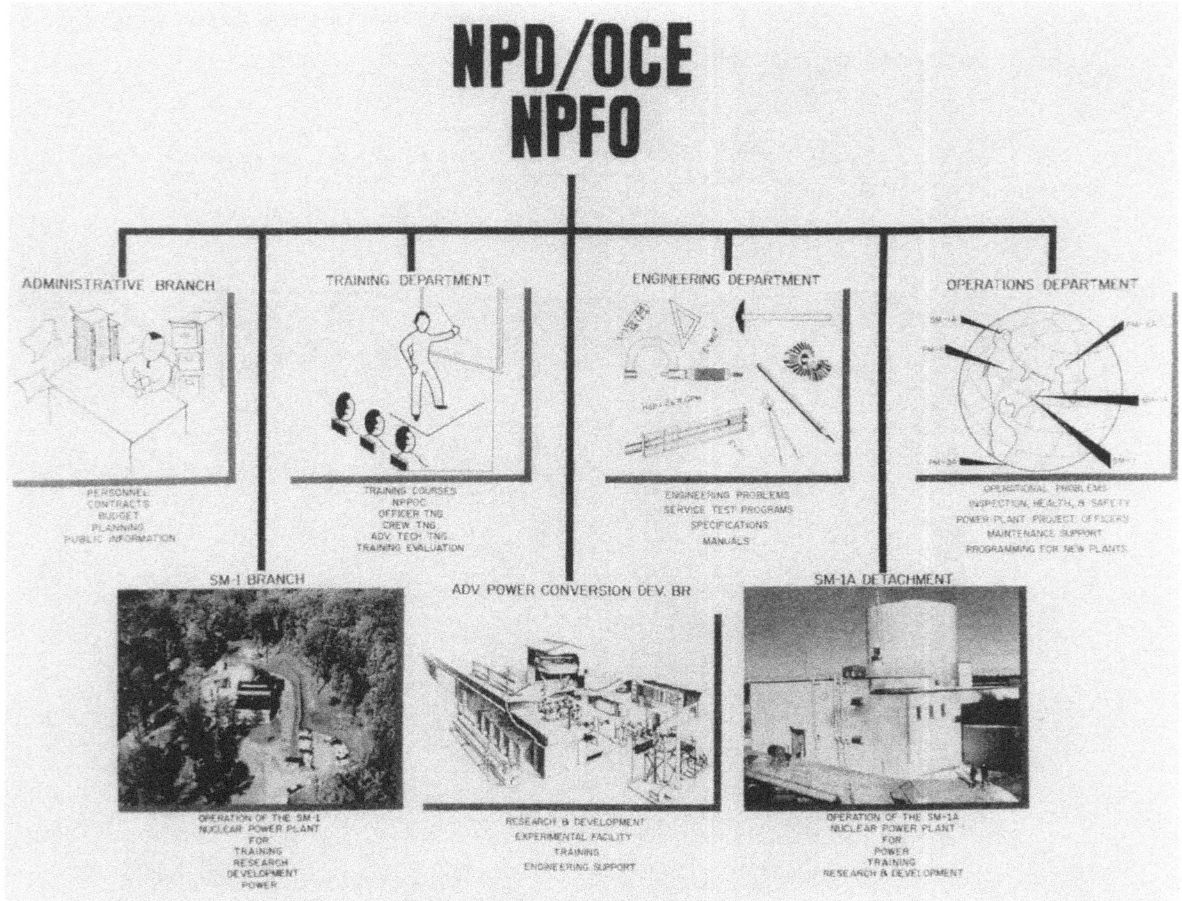


12. Army Corps of Engineers Nuclear Power Program organization chart.

CHAPTER 5

# THE NUCLEAR POWER PROGRAM IN FULL BLOOM

Although the SM–1A was the Army's first field facility to be planned and started, the delays in construction and testing prevented the Fort Greeley plant from becoming the first field reactor on the line. That honor went to the PM–2A, located in the ice tunnels of Camp Century, 138 miles inland from Thule Air Force Base on the Greenland ice cap and less than 900 miles from the North Pole. The Nuclear Power Field Office, U.S. Army Engineer Reactors Group, as the Army Reactors Group had been redesignated on July 15, 1958, began planning for the PM–2A in early October 1958.[1]

The requirement for the nuclear reactor grew out of the Army's desire to maintain a year-round research and development facility on the Greenland ice cap. During the 1950s, the Corps of Engineers had established several bases on the ice cap for military and scientific research. Camp Tuto, 14 miles inland from the Thule Air Force Base, served as headquarters for the Corps' Polar Research and Development Center and provided facilities for scientists conducting cold weather studies. At Camp Fistclinch, 220 miles east of Thule, the Engineers undertook an extensive investigation of Arctic construction techniques, including the use of snow as a building material. These bases operated only during the summer months, however, due to the difficulties of providing logistic support during the long arctic winters.

Two Air Force radar stations had operated year round on the Greenland ice cap until 1957 when they sank into the snow. They had demonstrated that continuous operation under severe arctic conditions was feasible, but the problems the Air Force had supplying them with fuel suggested that a nuclear power plant might provide the solution for maintaining a permanent research facility on the ice cap. As a result, when the Army began planning for the construction of a new base for research and development in Greenland, the Nuclear Power Field Office received the task of providing the power source.

The Corps of Engineers' experience in designing the SM–1A and negotiating the construction contract had shown that installation costs could run as high as six times that of a comparable plant in the United States. Not only were transportation costs higher, but the requirement for sending large numbers of highly skilled workers to a remote site greatly increased expenditures. Paradoxically, the high cost of labor resulted directly from the same environmental features that made nuclear power attractive. Premium rates had to be paid in order to attract persons of sufficiently high caliber who could properly perform the plant assembly within the stringent parameters of quality control. For the Engineers, the lesson quickly became apparent. Plants built in remote locations should be designed to require the minimum amount of on-site construction effort. To accomplish this, as much of the plant equipment as possible should be prefabricated in the United States where skilled labor could be obtained at reasonable rates and used more efficiently because the workers would have access to the full facilities of a manufacturing plant.

The ANPP (including NPFO, NPD/OCE, and Army Reactors Branch, AEC) had a second major input into its planning for a suitable nuclear reactor at Camp Century. In considering the possibility of building small atomic plants for two remote Air Force installations in Alaska during 1959, the service had desired to have the facilities completed within a single summer construction season. Although the plants were not built in Alaska due to budgetary considerations, the planning provided two additional criteria as guidelines for developing small nuclear reactors: (1) The reactor should be prefabricated to the maximum extent possible to minimize the time for construction at the site; and (2) To facilitate transportation, the plant should be preassembled into shipping units compatible with all modes of transportation, including airlift, and capable of being handled with routine construction equipment, which must meet the same requirement for transportability.

In anticipation of such requirements, in June 1958 the Corps awarded two design study contracts to alco and the Martin Company, the two nuclear contractors who, on their own, had been pursuing this design concept. Each received the requirement for designing a plant to produce 1,000 kilowatts plus 1 million BTU/hr of low-pressure steam. Their designs were to incorporate prefabrication to the extent that assembly in the field could be accomplished in the least possible time by the same troop crew who would later operate the plant. The shipping units were to be compatible with all modes of transportation including cargo sleds and C–130 aircraft.

The design for the Army's new base at Camp Century also provided inputs into the two design studies. The companies could draw on the data obtained from the earlier ice cap operations in anticipating the environmental conditions at the site. They also could prepare a specific plant layout in light of the information about the construction of the rest of the base.

Finally, they had a firm estimate of the electricity and low-pressure steam power requirements of the base.

In fact, the estimated power requirements provided by the Army turned out to be 50 percent higher than the design alco and Martin were developing. The increased requirement grew out of the Camp Century planners' decision to employ electricity instead of steam to heat the base. Since converting the reactor heat (in the form of steam) to electricity required a greater reactor capacity, ALCO and Martin tried to convince the base designers to change their minds and heat Camp Century with steam. The planners held to their original decision, however, on the grounds that electrical heating would be simpler to install and maintain than a steam distribution system.

In any case, during October 1958, the Nuclear Power Field Office began preparing contract specifications for a prepackaged nuclear power plant capable of meeting the requirements of the new Greenland base. The same month, the Chief of Engineers assigned overall responsibility for procurement and construction of the power plant to the U.S. Army Engineer District, Eastern Ocean. The District's experience in managing and supervising construction in Greenland, Iceland, northern Canada, Labrador, and Newfoundland would prove to be of particular value in working on a project in such a hostile environment.

In turn, on November 10, the District sent invitations to ALCO and Martin to submit bids for constructing Camp Century. The District preselected the two contractors because the time schedule for the base's completion required the successful contractor to have a design essentially completed so that he would be able to immediately proceed with the procurement of equipment whose manufacturing took a long time once the District had placed the order. With their designs so advanced, the two companies were in fact able to submit their bids on December 11. The District took only until January 23, 1959 to evaluate the proposals and negotiate the contract with ALCO.

The fixed-price, guaranteed performance contract of $3,226,560 required the contractor to fabricate and assemble the entire plant in the United States, perform a full-power operating test on the assembled unit using an oil-fired steam generator in lieu of the reactor component, and prepare the plant for shipment by May 20, 1960. Two operational unit price schedules were included in the negotiated contract. Under these, the Government could direct the contractor to exercise supervision over the installation and testing of the plant at Camp Century and be paid on a man-day basis for the services of his supervisory personnel.

An important provision of the contract was the Special Condition regarding liquidated damages. This was aimed at guaranteeing the availability of the plant equipment at the start of the construction season in Greenland. The contract clause provided for payments to the Government by the contractor of $1,000 per day beyond the scheduled delivery date of May 20,

1960. The $1,000-a-day rate would be paid for the first 41 days. For delays beyond this time, the daily rate of payment became $3,000. The Army established the two rates on the premise that if the plant was delivered during a 41-day period following May 20th, the construction schedule could still be met by airlifting all the plant equipment to Greenland. The $1,000-a-day payments to the Government would defray the additional cost of airlift. After 41 days, or July 1, the construction schedule could not be met and plant operation would have to be delayed a full year. The $3,000 rate would be used to reimburse the Government for costs of storing the plant over the winter and of providing another source of power for Camp Century for the 1961–1962 winter operation.

In addition to the contract with ALCO for the power plant equipment, the Eastern Ocean District awarded an Architect-Engineer contract on February 26, 1959 to Metcalf & Eddy Engineers for the necessary design and preparation of procurement specifications for the power plant buildings and utility services. The company received the contract as a result of its previous experience designing arctic facilities, including earlier Greenland bases. The District's Supply Division initiated the actual purchase of the materials to be used during construction of the power plant building at Camp Century. On its part, ALCO selected Peter F. Loftus Corporation as an Architect-Engineer subcontractor to complete the detailed design of the secondary system equipment.

To complete the procurement of equipment, assemble the plant, test all the components, and then package the plant for shipment within the allocated sixteen months required continual liaison between ALCO, the overall plant contractor, its subcontractors, and the several government agencies involved with the project. The work of each organization usually interacted with the progress of another party involved in building the Greenland facility. For example, ALCO's design and selection of material for procurement imposed a requirement for the Architect-Engineer of the base for space and facilities. In turn, both contractors had to look to the Army Engineers for much of the basic data on the design of snow structures and the performance of equipment under ice camp conditions.

Concurrently with the design and building of the power plant, work continued on the design of the undersnow base. The Army made the final decision on the exact location of Camp Century on May 17, 1959. Soon afterwards, the first construction troops headed for the location and on June 14, 1959, the soldiers began construction of the temporary camp to house men during the construction season. Once this work was finished, construction began on the base, using the Army's scheme of erecting simple buildings inside cut-and-cover snow tunnels. The method proved completely practical and by the end of the summer a good start had been made on the base. The main access tunnel was completed. Several side tunnels containing living quarters, the latrine, and water supply buildings were finished, and 300 feet of the main nuclear plant tunnel were completed.

In October 1959, the Training Branch of the Nuclear Power Field Office selected the men who were to make up the first crew of the PM–2A. At about the same time, the Office also chose the first crew for the SM–1A. Although every man in the Army's Nuclear Program was expected to be of the highest quality possible, even more care was taken in selecting men for the Greenland facility. The men would have to operate the plant after assembling and testing the reactor under conditions that offered little if any outside assistance. To facilitate their work, ALCO and the Training Branch prepared a special training curriculum as soon as the crew had been selected. Before leaving Fort Belvoir, the men received special education so that they would be familiar with the unique characteristics of the PM–2A when they reported to the ALCO factory to take part in the assembly and testing of the power plant before it was shipped to Greenland.

Despite numerous difficulties uncovered by the manufacturers of various components during construction, two of the skids on which the power plant was built had arrived at ALCO's plant in Dunkirk, New York, by early December 1959. At the same time, the first contingent of the crew arrived. Considerable fabrication of the plant, particularly of the primary reactor system skid, was still underway, so the crew was able to familiarize itself with all aspects of the systems and components. In addition, the crew was able to make suggestions to correct certain problems and recommend modification where potential difficulties became apparent.

Once the entire crew had arrived in Dunkirk by the middle of January, ALCO personnel began giving approximately 20 hours a week of additional instruction to complete the men's formal training. The plant assembly at ALCO paralleled as closely as possible the actual installation that would take place later that year at Camp Century; thus the crew learned what to expect when they and the power plant reached Greenland. Moreover, the assembly process revealed certain design problems that the contractor and crew could correct within the factory environment.

During March and April, the final assembly and performance testing took place. Although some components did not perform properly, the manufacturers were able to resolve the difficulties within the period allotted for testing. With this accomplished, the contractor and crew dissembled the plant and prepared it for shipping.

In light of the care and effort that ALCO and its subcontractors had expended on manufacturing the plant equipment, the company and the Nuclear Power Field Office made the same effort to prepare the plant components for shipment to Greenland. If any of the major components suffered damage, the completion and start-up of the power plant could have been seriously delayed due to the time required to manufacture replacements, the difficulty in delivering the new equipment if the delays ran into the winter months, and the problems encountered trying to accomplish repairs at the construction site. Consequently, ALCO's plans for packaging and shipment received careful review and each step of the procedure was inspected to ensure that each piece of equipment had sufficient protection.

To accomplish the actual packaging, ALCO awarded a subcontract to Page Airways of Rochester, New York, because the company had experience in preparing and shipping delicate equipment. As a result of the careful packaging only one ceramic top to a laboratory cabinet was damaged during transport. At the same time, due to careful inventorying of each crate as it was packed, including the photographing of the contents, the assembly of the plant went smoothly at Camp Century.

While final testing and packaging went on at the ALCO plant, the Army resumed work at Camp Century on April 17. It took almost a month to prepare the base for the construction troops. Although the aboveground camp had sustained little damage, snow had to be removed from the living quarters and the heavy equipment left on the site had to be located, removed from the snow, and made operational. By May 12, enough work had been accomplished to begin construction on the underground facilities. Immediate attention was directed toward the foundation and floor system of the nuclear plant tunnel started the previous summer and on the excavation of the remaining power plant tunnels. By the time the first skid of equipment arrived, all foundations except that for the reactor building were completed. As a result, the crews could begin immediately installing the power plant components.

At the close of the construction season the previous fall, the Army had determined that the construction schedule at the site could be met by delivering the plant equipment by sea to Thule on or about next July 10. The economics inherent in using surface transportation instead of airlifting the plant made the decision particularly attractive. Nevertheless, because the design of the PM–2A had been developed with the idea that the plant could be air transportable, the Army decided to fly the largest single skid of equipment to Greenland to demonstrate the practicality of the concept. The component chosen was one of the three air blast coolers, which measured 9′ × 28′3″ and weighed 30,000 pounds when packaged for shipment. It was also designated as the first skid to be placed in the main plant tunnel.

From Dunkirk, ALCO moved the cooler by semi-trailer to Dover Air Base, Delaware, and consigned to the Military Air Transport Service (MATS) cargo service as a routine shipment. MATS loaded the unit into a C–124 Globemaster, taking only 3 hours for the job. That evening, May 26, the plane left for Thule, making two intermediate refueling stops, and landing in Greenland thirty hours later. After being unloaded and trucked to Tuto, the jumping off point for Camp Century, the cooler was delivered by sled to the construction site and installed on its foundations by June 14.

The shipment demonstrated the validity of the concept of a portable nuclear power plant and provided the construction troops with an opportunity to practice handling a major unit prior to the arrival of the rest of the power plant by sea.

By this time, the Bristol Steel and Irons Works in Bristol, Virginia, the fabricator of the structural steel framework in the reactor building, had completed the trial erection of the structure as called for in its contract. In constructing the test exercise, the company trained the Army crew who would be doing the actual steel erection at Camp Century. When the trial fabrication ended, the steel framework and the erection crew were airlifted to Greenland with the expectation that the troops could begin work on the reactor building before the rest of the plant arrived. Due to construction delays, however, the foundation for the reactor room was not completed until the first shipment of power plant equipment arrived. The contractor had anticipated there might be delays in completing the reactor building, so the delivery schedule called for the primary system equipment and vapor container to arrive in the last swings. Consequently, the slow pace in finishing the reactor building did not affect the installation of the primary system.

In planning for the sea shipment of the power plant, ALCO and the Nuclear Power Field Office worked with the Military Traffic Management Agency (MTMA) to decide on the most suitable post and arrange for the special requirements needed to transport a nuclear power plant aboard ship. MTMA selected Buffalo as the port of embarkation because of its proximity to Dunkirk and its port loading facilities. The Military Sea Transport Service was directed to have a suitable ship available in time to be loaded and launched to Thule as the first cargo vessel of the 1960 shipping season. The ship selected, the USNS *Marine Fiddler*, had previously been used to transport locomotives. As a result, two of its holds had extra large hatches that permitted direct loading of cargo, which alleviated the need to slide crates once below deck.

As the second largest ship to navigate the St. Lawrence Seaway, the *Marine Fiddler* did represent some berthing problems in Buffalo. The day before it arrived, some dredging was done to remove a sandbar in the channel. A special train carried the power plant to Buffalo where loading was completed on June 25. The next day, the *Marine Fiddler* departed for Thule carrying the most valuable cargo ever shipped from the port of Buffalo. It arrived on July 10, the selected target date.

Once in the port, the Transportation Corps took over, unloaded the ship, and moved the power plant to Camp Tuto in three days of uninterrupted operations. At Tuto, a ramp road led up to the Greenland ice cap and the PM–2A was transferred from trailor trucks to cargo sleds for the journey to Camp Century. Some of the plant crew who had flown to Thule the previous week helped with the loading operation while the rest of the men went directly to the base to begin preparations for the installation of the plant.

The first heavy swing of power plant equipment left Tuto at noon on July 13 and arrived at 4 a.m. on July 17. The Transportation Corps had marshaled a sufficient amount of trail equipment and was able to begin mounting a second cargo swing almost immediately. Following on the path broken by the first swing, the second swing arrived just before midnight on July 17. The arrival of the two swings the same day amounted to a considerable improvement over the scheduled delivery of the plant components. This ensured an ample amount of equipment on hand to permit work to progress as rapidly as possible even before the third swing carrying the balance of the equipment arrived from Tuto on July 28.

The construction crews benefited from the continuous daylight of the northern summer. Temperatures in the tunnels averaged 10°F and the wind was never strong enough to cause serious drifting of snow. If high winds had occurred after the excavation of one of the tunnels and before it could be covered, snow might have refilled the trench. Drifting snow would not have supported the heavy excavation snow plow and hand labor would have been needed to clear the tunnel. As a result, the construction crew had anticipated that it would have been easier to simply dig a new tunnel than reconstruct one already filled in. Fortunately, this problem never occurred.

Instead, within the first 4 days of the arrival of the initial two swings of equipment, the construction troops had emplaced five skids containing two air blast coolers, the heat exchanger, condenser, and turbine generator. As soon as the skids were in place, the plant crew began installation of the interconnections between the equipment. The men were divided into two groups each working a 10-hour shift. They received needed supervision from ALCO personnel on-site and the installation progressed rapidly, without much difficulty. One of the few problems occurred when an air blast cooler was emplaced backwards due to an error in exterior markings. The plant crew required only seven hours to rotate the unit to the right direction.

By September 16, the last of the equipment had been installed in the reactor building, and the rear wall was closed, completing the structure. The snow shield wall separating the reactor building from the hot waste building could then be started. With this done, work began on the hot waste building. When its floor system was completed, the hot waste tank, the last piece of equipment, was put in place and the reactor tunnel was closed over. The hot waste building was completed on September 25.

In the meantime, by August 12, all piping had been installed throughout the plant as far as the vapor container. By the same date, all the secondary system wiring was in place except for the connections from the generator to the switchgear. The only real difficulty that arose during the installation of the electrical cables occurred inside the vapor container when it became necessary to rewire the junction box to provide proper circuit connections with the control skid.

Delays in manufacturing the building panels did prevent a start on the building erection as originally scheduled. The plan had called for panel erection to begin as soon as the first skid was in place and the interconnecting piping was installed in the area. Instead, panel erection did not start until August 19. However, even when work did finally begin, erection proceeded slowly because of dimensional errors and problems sealing interior joints between the panels. A three-week delay in beginning panel erection meant the same delay in having heated enclosures around the plant equipment. As a result, men had to do more work in cold tunnels than expected. In addition, this meant a delay in some of the mechanical testing that required filling systems with water.

Once the construction crew did complete a portion of the plant buildings, however, electrical crews installed the permanent electrical heaters so that the plant crews could warm the equipment. Temporary heaters were also used to warm the partially completed buildings. Nevertheless, pipes near the floor were constantly in danger of freezing and on occasion did freeze. In any event, the plant interconnection work proceeded on schedule.

By mid-September, however, trouble had developed in trying to complete the plant's utility systems. Although two of the major systems, fire suppression and ventilation, had been prefabricated in the same way as the plant equipment, the other utility systems—heating, fire alarm, water, sewerage, and manometer—had to be built in place from essentially standard materials. The Army construction troops were nearing the end of their maximum temporary duty of six months, and as they left fewer men were available to work on the utility installation. Moreover, most of the on-site construction personnel had only recently graduated from specialized training courses and so had little practical experience in their trades. This slowed some of their work on the utility systems. Although the Army did replace some of the departing men, it became obvious that more experienced men would be needed if Camp Century was to be completed on schedule.

To solve the problem, the Eastern Ocean District negotiated with Greenland Contractors in Thule and reached an agreement to have nineteen men and a superintendent come to the construction site the last week in September to assist on installation of the utility systems. The group included carpenters, electricians, pipe fitters, sheet metal workers, heating and ventilating specialists, and ironworkers. While the integrity of the troop and the contractor crews was maintained by assigning complete and independent jobs to each crew, the men found themselves working side by side using the same equipment. Ultimately, therefore, the two work forces essentially became one.

With progress came a series of minor problems and modifications of plans.

In most cases, the construction crew could fabricate new parts or correct design flaws in the equipment or building components. The additions and omissions demonstrated that even with a blueprint as detailed as the PM–2A and a preassembly of most of the plant equipment, unexpected problems would likely crop up. To a significant extent, because the testing of equipment began concurrently with the plant installation, potentially troublesome flaws were recognized and corrected before causing serious operating trouble.

The plant crew checked each piece of equipment as they installed it. They pressure tested each piping system or major portion of it as it was completed. Each cable system received both insulation and continuity checks. Each piece of operating equipment was operated and inspected for performance. The plant utility systems underwent the same testing. In addition to catching flaws before they caused trouble, the simultaneous testing during construction had the advantage of allowing time for replacement or repair of the defective component before the scheduled start up of the plant.

One example of how early and continuous testing of equipment saved time was the process by which the crews found and repaired leaks in the vapor container caused by an error in manufacture. Welding did not work and repair equipment had to be ordered and brought to Camp Century. But while waiting for the arrival of the order, the crew was able to accomplish the initial operation of the reactor at zero power, among other jobs. Although the problem delayed completion of the plant, the time table could have been more seriously affected if testing had waited until just before the planned power operation.

In any event, the simultaneous construction/operational testing process makes it difficult to select a specific completion date for the PM–2A. Nevertheless, October 1 probably marks the end of construction and the beginning of the final plant testing because core loading operations began on that day. At 10:30 p.m. the spent fuel rack was tested to check for subcriticality of the rack when loaded under the most reactive conditions. Even when the rack was fully loaded with fresh rods and with the spare fuel elements added and the absorbers removed, the assembly remained subcritical. This initial phase of the testing ended at 3 a.m. on October 3.

The crew then began loading the core into the reactor vessel. When it had completed the job, the crew sealed the reactor and all personnel evacuated the reactor building and adjacent tunnel. Controls were locked in the zero power mode. At 6:52 a.m., the control room operator brought the reactor to criticality. After five minutes, the operating crew manually scrammed the reactor. The leaks in the vapor container had not yet been repaired; thus the reactor could only be operated and tested in the zero power mode. Finally, on November 8, with the repairs completed and the integrity of the vapor container established, the primary system heat up began.

Problems began to manifest themselves immediately. During the night of October 8, two gaskets on the secondary diversion line to the hot waste tank blew out. After the crew replaced these and heat up began again on October 9, a trap overflowed and water backed into the first stage of the turbine. This is turn damaged the main rear turbine bearing. When the crew repaired this problem and restarted the turbine, the generator would not accept a load because of a misconnection in a relay. Once the crew corrected this error, the power plant went back on the line and easily carried 250 kilowatts of the base's power load. However, the crew found the reactor would not work in tandem with the standby diesels. Ultimately, the crew traced the trouble to a too-high water level in the steam generator. On November 12, 1960 at 5:30 a.m., with the problem corrected, the PM–2A carried the Camp Century power load both separately and in parallel with the standby generators. Thus, the reactor became the first nuclear power plant to supply power and heat to an isolated military base.

Bugs continued to manifest themselves, however. The crew soon discovered that radiation levels registered higher than design limits in areas of required access to the reactor tunnel and building. Following the first power run, another radiation survey confirmed that radiation doses were unacceptably high. Unfortunately, immediately after the plant crew transmitted the first sketchy reports of the survey, adverse atmospheric conditions caused a 10-day radio blackout.

During this time, the plant crew checked everything that could possibly contribute to the high dose rate. Some relief was obtained by filling the dry cap over the top of the vapor vessel. And, because dose rates in all areas of the power plant were acceptable except in the reactor tunnel, the crew gave consideration to operating the reactor through the winter without adding any additional shielding. Ultimately, it was determined that the reactor should not be operated until the shielding was improved, except as necessary to generate data needed in the redesign process.

By the first week in December, it became obvious that the solution to the shielding problem would require enough time to justify returning most of ALCO's personnel to the United States. In this way they could assist in interpreting the radiation test data and contribute their first-hand knowledge of site conditions in designing the necessary modifications. Even before the men returned, however, work had begun in ALCO's Nuclear Power Engineering Department in Schenectady to determine the source and solution to the shielding problem. They faced two difficulties. First, new shielding had to lower the dose rate on the upper level of the reactor building during reactor operation. Second, it also had to lower the dose rate inside the vapor container following shutdown.

Representatives of the Eastern Ocean District and the Army Nuclear Power Field Office worked with ALCO's design engineers to produce a plan that would meet three guidelines established by the District Engineer. The modifications to the shielding, if not permanent, had to permit the plant to operate safely until conditions favored the installation of a permanent solution. Moreover, any shielding installed could not prevent access for further work if the initial attempt at corrections did not prove satisfactory. Finally, corrective action had to be taken immediately to provide a workable nuclear power source for use through the rest of the winter at Camp Century.

To handle the problem of shutdown dose rate inside the vapor container, the engineers mandated 4 inches of lead brick on the surface of the primary shield tank. To stop the high dose rate on the upper platform of the reactor building, the engineers applied a combination of shield devices. They also decided to spray the snow walls of the reactor tunnel with a water solution of boric acid to decrease absorption of neutron radiation and its subsequent emission of gamma radiation.

The pig lead was the first of the radiation shield materials to reach Camp Century; it arrived on December 10 along with the steel needed to support the bricks. The shielding materials were delivered to the base by a combination of heavy swings operating at maximum capacity, and air drops from an Air Force Air Rescue C-47 based at Thule. By January 4, 1961, when ALCO personnel and the Eastern Ocean District Resident Engineer returned to Camp Century, the construction crew had completed melting the pig lead into 2-inch-thick bricks and positioned them 1-row thick against the primary shield tank.

The weight of the new shielding was approaching the point at which it would become necessary to modify the reactor building structure to support another row of lead bricks. Consequently, the Engineers decided to stop at a 2-inch thickness until they could check its effectiveness by operating the reactor. Ultimately, the single layer proved adequate, which saved 10,000 pounds of weight. In the meantime, the crew implemented the other planned measures and completed the shielding modifications on February 5.

The operating crew spent the next day inspecting the plant equipment and reviewing the PM–2A's operational procedures for the new members who had reported to Camp Century since the first start up. Finally, on February 7, the crew brought the reactor to criticality, but held it at low power in order to perform the remaining contract tests. During the next two days of operation at various levels, the crew checked the data, made necessary adjustments, conducted a new radiation survey, and corrected a number of minor problems.

Finally, just before midnight on February 9, the crew started the 400-hour performance test with the PM–2A carrying the total required power load of approximately 680 kilowatts. Following the completion of the test on February 26, the crew devoted a week to inspecting the plant equipment and making minor repairs. With everything in order and the radiation problem cured, the Polar Research and Development Center accepted the plant from ALCO.

The plant had required seventy-seven days to assemble at Camp Century. It broke new ground in many ways, including its modular construction of the components and their shipment on pre-assembled skids. Other new features included the air-blast coolers for condensing turbine exhaust steam and the Rodriguez Well, which provided the base's water supply by using steam from the nuclear power plant to melt ice from the 6,000-foot-thick Greenland ice cap.

Many of the start-up problems of the PM–2A undoubtedly stemmed from the tight design and construction schedule and the new technology used to create a truly portable nuclear power-generating system. Nevertheless, many of the problems with the facility, even after the Army accepted the plant from ALCO, had to do not with the nuclear system, but with conventional aspects of the operation. From August 2 to November 7, for example, the PM–2A was shut down to make repairs to the turbine generator that had been causing problems for several months.[2]

Once the plant crew had realigned the turbine and installed new oil-cleaning equipment, the plant entered a period of dependable service. During March and April 1962, the PM–2A stayed on the line 99.7 percent of the time. And on May 8, it ended a record run of 864 hours. The plant shutdown did not result from any mechanical problems but from the settling of the snow arches over the tunnels onto the plant roof. After ascertaining the extent of the situation, the PM–2A went back into service until June 13, when it was shut down to rebuild the arches. After the repairs were made, the plant went back on the line on August 19 and proceeded to set several records for continuous operation, including one of 1,038 hours and later of 2,502 hours.[3]

During this period, the PM–2A demonstrated the validity of the Army's Nuclear Power Program by providing dependable power in a remote area unencumbered by the logistic considerations inherent with conventional power plants. A diesel-powered plant providing an amount of power equivalent to the PM–2A would have used some 400,000 gallons of fuel per year. Nevertheless, the nuclear plant was not able to justify its existence in Greenland in terms of cost-effectiveness. Although the plant had been designed to provide 1,560 kilowatts of power to meet Camp Century's expected energy needs, once in operation the base required only 300 to 500 kilowatts of power. Moreover, by the spring of 1963, even this power requirement began to decline as the Army's polar research and development activities at Camp Century began to be phased out. At the same time, the snow arches over the tunnels again threatened the integrity and existence of the base.

By July, the plastic condition of the tunnels, aggravated by heat generated within the undersnow camp, required constant trimming to prevent buildings and utilities from being crushed. In the area of the reactor, several tunnels already had been rebuilt. As a result, the Army Materiel Command (AMC), to which activity the Polar Research and Development Center had been transferred in 1962, decided to operate Camp Century only during the summer season. Because the standby diesel power plant could supply the limited requirement of the facility on such a seasonal basis, the PM–2A was no longer economically viable. Therefore, in late July, while the reactor was shut down for planned maintenance, the Army announced that the PM–2A would not resume operation. This gave the Nuclear Power program the opportunity to demonstrate the feasibility of relocating a portable nuclear plant.

The Nuclear Power Field Office had already anticipated the possibility of having to relocate a portable nuclear plant and, in mid–1962, had two former officers-in-charge of the PM–2A analyze the task of moving the facility based on their experience installing and operating the plant. In their report, they noted that the transfer of the primary system would pose a major difficulty because of residual radiation. Subsequently, the Field Office contracted with Burns and Roe, an architect-engineer firm, to study the feasibility of dismantling and relocating the power plant.

The company focused a major portion of its study on determining radiation levels in the system at intervals of 3 and 12 months after a shutdown so that dismantling could be safely undertaken. From its research, Burns and Roe concluded that it would take 12 months for short-term residual radioactivity to disappear. This period coincided with a projected two-season dismantling effort. During the first summer, plans called for the plant to be shut down and the secondary system removed. The next summer, it was believed that the primary system could be removed. Nevertheless, because of the hazards of dismantling the primary system with its radioactive components, Burns and Roe recommended shipping the entire vapor container with both skids in a single package. The total weight of 100 tons made the proposal impractical because there was no crane within 2,000 miles of Thule that could handle such a load. Consequently, the Nuclear Power Field Office set out to prepare an alternative relocation plan.

As formulated, the plan incorporated certain aspects of the Burns and Roe study along with the experience of the Nuclear Power Field Office. The Office assigned both the U.S. Army Research Support Group and the U.S. Army Engineer Reactor Group, both based at Fort Belvoir, the task of removing the PM–2A. Using technicians from the Nuclear Power Field Office and an Engineer construction platoon, the organization prepared a plan for removing and transporting the reactor and other components. Two factors dominated the planning. The exposure of troops to high radiation levels and the discharge of radioactive materials had to be avoided. The plan had to keep in mind the short arctic shipping season that lasted only from May to September.

Once the Army made the decision to deactivate the PM–2A, the Nuclear Power Field Office set the plan in motion. With the reactor shut down on July 9 for regularly scheduled maintenance, the plant crew began work on

August 1, to "winterize" the plant by draining all except the primary system and removing all components that could be damaged by freezing. Once this was completed by mid-August, they began planning for unloading the spent fuel from the core. Initially, the crew expected to complete this phase of the operation in early September, but September 27 proved to be the earliest possible date for safe shipment. In preparation for this, the crew removed the dry-cap cover, steam dome, primary vessel cover, and the water in the spent fuel tank. The primary system was then circulated through demineralizers and filters to remove radioactive waste.[4]

To unload the reactor itself, the crew removed the thirty-seven fuel elements one by one and placed them in the spent-fuel tank. The elements were highly radioactive, so they had to be shielded during transport. To do this, the crew received 7 steel shipping casks, each lined with a 10-inch layer of lead shielding. To lower and then raise each of the 10-ton casks, the 864th Engineer Battalion erected a 70-foot double–single Bailey bridge with a triple-story center section over the reactor trench. The center section contained a 10-ton electric overhead traveling hoist.

During the removal operation, the casks were lowered by the hoist through hatches into the reactor building and finally down to the floor of the spent-fuel tank. The crew then used special, remote-actuated tools to transfer the elements from the spent-fuel tank to the shipping casks, which were then sealed, washed down with detergent, hoisted back to the surface, and in a few days transported by heavy swing to Camp Tuto for storage until the 1964 shipping season. The water in the spent fuel tank, primary system, and all shield tanks was circulated through the demineralizer and filters to reduce radioactivity to acceptable levels before being discharged into the hot waste well. Instruments susceptible to damage from low temperatures were packed and shipped to Camp Tuto for inside storage through the winter. Other instruments were boxed and left at the base. The crew completed the winterizing of the PM–2A on October 7 and left the next day. Two weeks later, the Army shut down Camp Century for the winter.[5]

The Army reopened Camp Century the next April and on the return trips from the base, the heavy transport swings brought back the packaged PM–2A skids and other equipment. The final components left Camp Century six weeks ahead of schedule on June 30, and arrived at Camp Tuto on July 4. From there, all equipment was trucked to Thule to await the arrival of transport ships. After arrival and loading, the USNS *Greenville Victory* left Greenland on July 27 and the USNS *Robinson* left on August 1, arriving at Hampton Roads Army Terminal on August 4 and 16 respectively.[6]

Although the Chief of Engineers advised the Air Force, the Navy, and Major Army Commands that the PM–2A was available for relocation, none of the services expressed an interest in acquiring the plant. As a result, the radioactive primary system, consisting of fifteen pieces, was loaded on 8 flatcars and shipped to the National Reactor Testing Station in Idaho. The radioactive-free secondary system, consisting of thirty-four pieces, was shipped on fifteen flatcars to New Cumberland Army Depot, Pennsylvania. The 3 vapor container sections were off-loaded onto a barge and sent to the Savannah River Nuclear Facility via the Intercoastal Waterway.[7]

While the active life of the PM–2A lasted less than three years, its construction, operation, and removal demonstrated the feasibility of small, portable, nuclear power plants. Dr. C. F. Jacobsen, who represented the Danish Atomic Energy Commission during the removal of the PM–2A, provided an outsider's judgment of the significance of the plant. In a letter to Capt. Joseph Franklin, Officer-in-Charge during the removal of the PM–2A, Jacobsen wrote:

> I have emphasized the extremely high quality of the job done under adverse conditions, both the running of the reactor and the dismantling and the clearance. Actually, I have proposed that we use your work as propaganda for atomic energy, as you have proven that reactors can be used and taken apart without danger to anybody even without the use of highly specialized equipment which one would use if a permanent plant should be dismantled. It shows that highly qualified leaders with a sense for "good housekeeping" is what is needed. Maybe it will help in removing the anxiety about nuclear reactors, which corresponds to the demand that a man with a red flag walked in front of the earlier railway trains.[8]

The contribution of the PM–2A to the advancement of nuclear energy did not end with its removal and storage. From April to November 1966, Phillips Petroleum Company, as contractor for the Atomic Energy Commission, ran a series of tests on the primary system of the PM–2A at the National Reactor Testing Station. These tests were part of the AEC's reactor safety program and concerned the neutron embrittlement of carbon steel. One problem that had been facing all nuclear reactors was the fact that irradiated steel, such as that used in a pressure vessel, loses its strength over a period of time. Using the PM–2A's pressure vessel, tests were performed from −20°F to 272°F, at pressures between 0 and 5,250 pounds per square inch. No brittle fracture occurred even when the testers gradually machined a defect into the vessel. A fracture finally occurred at −20°F and 4,475 pounds per square inch pressure after hydrochloric acid was introduced into the machined defect.[9]

That it took such an extreme test, ultimately using an artificially induced defect, demonstrated the high quality of the PM–2A pressure vessel. Ironically, during its development, some AEC engineers criticized the design as being marginal, if not inadequate, in its ability to withstand neutron embrittlement for less than one core life. The test reflected favorably on the competence of ALCO and the Corps of Engineer reactor design teams and provided the nuclear industry with important information about the embrittlement problem.

In the meantime, the Army's Nuclear Program continued to expand.

Although the PM–2A became the first portable reactor to begin producing power, it was not the first one on which the Army had started work. That honor went to the PM–1, which was initiated on November 9, 1957, by Air Force request to the AEC. In its letter, the service asked the AEC to design, develop, fabricate, erect, and test operate a nuclear power plant at a remote facility. The AEC, the Air Defense Command, and the Air Force Systems Command jointly selected the Sundance Air Force Radar Station 6 miles from Sundance, Wyoming, in the northeast corner of the state.[10] The Air Force approved this locale.

Whereas the station was reasonably accessible, it nevertheless fit many of the criteria of a remote base of the type the Army's Program had envisioned as suitable for small nuclear power plants. The site was located at an elevation of 6,650 feet above sea level on Mount Warren, in an area far from any significant population centers. It was subject to severe weather conditions with the mean average snowfall of 110 inches and temperatures that on occasion fell lower than −40°F. In addition, water supply was limited.

As its primary purpose, the PM–1 was to provide electric power and space heating for the 731st Radar Squadron of the North American Air Defense Command (NORAD). Its secondary mission was to yield engineering and operating data that could be used to design improved nuclear power plants for Distant Early Warning (DEW) line and other isolated Air Force stations. To accomplish these goals, in March 1959 the AEC and the Air Force awarded a jointly funded contract to the Martin Company of Baltimore. The contract called for designing, fabricating, packaging, installing, and initially operating and testing the PM–1. Under the terms of the contract, the Martin Company was also required to conduct a training program for the military personnel initially assigned to the power plant with the objective of developing a crew with complete operation and maintenance capability.

In October 1959, the Martin Company submitted to the AEC the initial PM–1 Nuclear Power Plants Hazards Summary Evaluation. In addition to the evaluation, the report described the preliminary design of the plant. The AEC and the Air Force approved the design, which was essentially followed through to the final blueprint. In February 1961, the company submitted an addendum to the original evaluation, which contained the detailed description of the final design of the PM–1.

The plan called for a facility to generate 1,000 kilowatts of power for normal station use and radar operation as well as delivering 7,000 pounds per hour of low-pressure steam for space heating. To produce this power, the PM–1 would use a fuel core about the size of an oil drum that would be expected to operate for two years without refueling. Unlike the other military plants, the PM–1 was to have no containment vessel. Instead, to minimize the release of radioactive material and to isolate potential contamination areas, the Martin design placed the primary system in three underground storage tanks beneath the reactor building. Although the tanks lacked the strength of a containment vessel in the event of a major accident, the shield water, the ground above the tanks, and the primary building itself would limit the dispersal of radioactive material. Moreover, given the remote location of the plant, even the worst possible accident would not create a civil disaster.

Once the Air Force and the AEC had approved the design, the Martin Company fabricated the plant's components at its plant in Baltimore. As with the PM–2A, the contractor mounted all components except for the reactor on skids and assembled the nonnuclear portion of the plant in its Baltimore factory. After receiving their training at the Nuclear Power Field Office at Fort Belvoir, the PM–1's initial operating crew, consisting of 18 men from the Air Force, 2 from the Army, and 2 from the Navy, assisted Martin personnel in the assembly, testing, and disassembly of the power plant. Meanwhile, on Mount Warren, construction began to prepare the foundations, buildings, and supporting structures for the arrival of the plant.

In June 1961, after testing was completed, the components were packaged in modules limited to outside dimensions of 8′8″ × 8′8″ × 30′ so that they would fit within a C–130 Globemaster. Martin then shipped the 16 packages the 1,700 miles from Baltimore to the Sundance Station. Working under the supervision of company engineers, the military crew installed the plant. Following its completion on December 16, the crew took two months to run the necessary series of pre-critical tests on the plant systems. On February 25, 1962 they loaded the core and the PM–1 attained criticality.[11]

The power plant then underwent its break-in period. During testing, the contractor and crew determined that design specifications for power output could not be met for sustained operation. To correct the problem, Martin had to replace the air ejectors with larger units, replace the transformer with a larger unit, replace the solenoid-operated valve in the primary system with an air-operated valve, and modify the shield water cooling system. After its shutdown from August 8 to 31 for the modifications, the PM–1 embarked on a period of reliable operation. After completing its required 400-hour performance test on October 1 to satisfy the AEC testing program, the plant was turned over to the Air Force on October 31.[12]

As with the SM–1A and the PM–2A, the PM–1 did require an extended shakedown to correct several design and mechanical defects. Although the Air Force maintained ownership of the plant, the service's Air Defense Command entered into a technical support agreement with the Corps of Engineers to handle technical support on the PM–1. The Air Defense Command agreed to using the Engineer Reactor Group as its technical base for the plant and in return, it agreed to provide a project officer and funds for use at the Nuclear Power Field Office.[13]

Despite several shutdowns during 1963, including a two-month shutdown to repair a control rod system malfunction, the PM–1 still had several long power runs. In April it had a sustained run of 510 hours. Beginning November 24, the plant began a record power run of 2,630 hours, which did not end until April 1964. During October 1964, the plant stayed on-line 83 percent of the time; in November, 94 percent of the time; and in December, 99 percent of the time. Before undergoing its first refueling in August 1966, the PM–1 broke the SM–1A record of 2,750 hours of continuous operation. In so doing, the plant confirmed the reliability of the Army's small nuclear power plant program for remote areas. Nevertheless, it remained for the PM–3A, the Navy's first land reactor (located at McMurdo Sound, Antarctica) to operate in total isolation for an extended period of time.

During the first International Geophysical Year from July 1, 1957 to December 31, 1958, 11 nations established a total of fifty-six scientific research stations in the antarctic and subantarctic regions. The Navy received the assignment for providing logistical support to U.S. participants in the international research program. For this purpose, the Navy set up 7 research stations. The station on Ross Island in McMurdo Sound served as the logistical base and center of operations for supplying the other bases.[14]

When the International Geophysical Year ended, the U.S. antarctic research program continued its study of the antarctic region. While Byrd Station, Ellsworth Station, and South Pole Station remained in operation, the Navy Air Facility at McMurdo Sound stayed at the center of activity. The base, built on solid ground, 80 feet above sea level, contained a fully equipped biological laboratory and a small library of science references. Although climatic conditions were milder on Ross Island than in the interior of Antarctica, they were still very severe. The lowest official temperature recorded at the base in August 1958, was −59°F., with winds clocked at 85 knots.

From the beginning, logistical support for the Antarctica stations proved to be a sizable undertaking. The severity of the winter limited shipments to the milder months of October through February. Supplies and personnel went from the United States to McMurdo Sound by sea or air and from there were transhipped as necessary to the outlying stations. In 1960, about 70 percent of the cargo consisted of some type of fuel for the diesel power generators, for cooking and heating, and for the transport of aircraft.

In the search for ways to reduce the logistical hardship of bringing fuel oil to Antarctica, the Navy turned to nuclear power. In early 1960, Kaiser Engineers and Contractors, Inc., submitted a study to the AEC comparing the cost of generating nuclear and conventional power. Relatively speaking, nuclear power proved to be far more cost-effective than conventional power at the Antarctica site as well as at other sites—including Okinawa, Guam, and Thule Air Force Base. The study also recommended that for McMurdo Sound, the Navy should use a factory-assembled, air-transportable, pressurized water reactor of the type developed for Camp Century and Sundance, Wyoming.

This report and other studies confirmed to the AEC and the Congressional Joint Committee on Atomic Energy that Antarctica would be a proper and economically viable site for applying nuclear power to a remote military facility. As a result, the Navy asked the AEC to develop a plant suitable for operation at McMurdo Station under the supervision of the Army's Nuclear Power Program. As part of its request, on July 6, 1960, the Secretary of the Navy established the U.S. Naval Nuclear Power Unit at Fort Belvoir under the Navy's Bureau of Yards and Docks. The Agency had the responsibility for the design, procurement, operation, and maintenance of nuclear shore systems and for the training of personnel to support them.

Following up on its hearings, the AEC's recommendations and the Navy's actions, in August Congress authorized the building of a nuclear plant at McMurdo Sound. In turn, the AEC awarded a fixed-price contract to the Martin Company to design and fabricate a 1,500-kilowatt reactor similar to the PM–1, which the company was building for the Sundance Air Force Station. The contract specified that Martin would supervise installation and initial testing and operation of the plant. In addition, Martin agreed that if the power plant, designated the PM–3A, failed to deliver power as promised, the company would reimburse the Government for the fuel oil it would have to burn to replace the anticipated power from the nuclear reaction.

Like the other portable nuclear plants already under construction, the PM–3A would have to be assembled during the short, relatively mild southern summer season. Martin therefore had only from August 1960 to November 1961 to fabricate the plant, assemble, and test it in its factory. With this done, the company would have to disassemble it, package it for shipment, and insure delivery to McMurdo Sound for erection before the antarctic winter began. Moreover, the contractor faced design and operation restrictions imposed by the 12-nation Antarctic Treaty, which prohibited the disposal of radioactive waste on the subcontinent.

As soon as the contract had been signed, the Navy's Bureau of Yards and Docks began working with the Martin Company to prepare designs for foundations, buildings, and supporting structures at McMurdo Sound. A joint team from Martin and the Nuclear Power Field Office travelled to Antarctica that fall and selected a site for the reactor about a half mile southeast of the main camp on the side of Observatory Hill overlooking McMurdo Sound. In December, ships brought in foundation materials and construction equipment which the Navy Seabees of Mobile Construction Battalion One used to prepare the site. Following the winter season, the Seabees began work the next October to erect the prefabricated, insulated, steel buildings which would house the PM–3A.

Meanwhile, the Nuclear Power Field Office used the same procedures to train Navy personnel to man the PM–3A as it had developed during the training of the operating crews for the PM–1 and PM–2A. Twenty Navy men went through the Nuclear Power Plant Operators Courses during 1960 and 1961. They then went to the Martin plant in Baltimore in May 1961 to participate in the pre-shipment assembly, testing, and evaluation of the PM–3A. With these tasks completed, the crew spent October working with the contractor's personnel packaging the plant. It was then shipped by land to Davisville, Rhode Island, the home of the U.S. Naval Construction Battalion Center, through which supplies and equipment for the PM–3A would subsequently be processed. After loading, the USS *Arneb* left for McMurdo Sound on November 1 and arrived on December 12.

The first operating crew had already arrived at the end of November and, with the Seabees, unloaded and moved the plant's components and equipment by tractor-drawn sled across the ice and up Observatory Hill. Martin personnel, the operating crew, and the Seabees, under the supervision of the AEC's New York operations office, then assembled the PM–3A in 71 days. The reactor initially went critical on March 3, 1962. Following a period of extensive systems shakedown and testing, it began supplying usable power to the McMurdo Station on July 10, 1962, giving the Army's Nuclear Power Program five on-line reactors.

Like the PM–1, the PM–3A was built above ground except for the primary system, which was placed in underground tanks. The only significant difference between the Sundance plant and the McMurdo facility was the addition of a fourth containment vessel, which was included for emergency purposes to handle expansion and prevent a radioactive leak into the environment. To fulfill the requirement of the Antarctic Treaty, the plant used a liquid waste concentration system similar to that of the PM–1. In the case of the PM–3A, the highly concentrated sludge was sealed in concrete-lined drums and buried at sea. Likewise, gaseous waste was compressed into cylinders and disposed at sea.

For the next 20 months, navy crews operated the plant under the contractor's direction. During this period, the men ran tests and eliminated troubles arising in the control rod drive mechanisms, nuclear instrumentation, hydrogen generation, and radioactive waste disposal. Finally on March 12, 1964, the Navy accepted the plant from the Martin Company. In doing this, the Army's Nuclear Power Program again demonstrated the feasibility of building and operating portable nuclear power plants in remote locations.

Despite the severity of the climate at McMurdo Sound, which lowered the efficiency of construction workers by 50 percent and slowed the handling and movement of materials, the plant was completed and went critical during one construction season. Although the PM–2A operated under similar environmental conditions, the men at Camp Century did have limited access to the outside world in the case of emergency. In contrast, in total isolation, the PM–3A had to supply power from March through October for the McMurdo base and to the population of 150 servicemen and civilians who remained in Antarctica.

To operate the PM–3A under these circumstances required a crew of 2 officers and 23 enlisted men. The three services furnished members of the crews in order to broaden their experience with different types of power plants. Each enlisted crew member had to attend the basic one-year nuclear power operators course at Fort Belvoir and qualify at another pressurized-water nuclear power plant in the specialty he was to perform on the PM–3A. Because of the lack of access to McMurdo Sound during the winter months, the crews were replaced in a unit during the summer months from October to February. Replacement crews were formed and trained as a team at the Nuclear Power Unit at Fort Belvoir. After reaching McMurdo Station, a relief crew received additional training from the crew already there until each new member attained the required proficiency. At the end of the first month the old crew began to phase out and, by the end of the second month, the new crew was in place.

The Navy Nuclear Power Unit was responsible for engineering and technical support for the PM–3A and helped train crews at Fort Belvoir. The Unit normally communicated with McMurdo Sound by radio-teletype over the Navy's communication system. Mail service could only be maintained during the summer months and the base sometimes would go for a month or more without voice radio because of atmospheric conditions.

Under these circumstances, reliability was crucial; the PM–3A was able to supply the needs of the Antarctic facility. On August 3, 1963, the plant completed a continuous power run of 1,134 hours. On May 6, 1966, it had a run of 2,002 hours and on October 9 of the same year, it completed a run of 3,390 hours—at that time a new record for any military power plant. Moreover, from July 10, 1962 to the end of 1972, the PM–3A supplied 60,400,800 kilowatt-hours of power to the McMurdo Station, the equivalent of 4,480,771 gallons of diesel fuel.

If the PM–3A and its four sister plants demonstrated their reliability and potential for cutting logistics problems at remote bases, the Army's Program had its drawbacks. In particular, the cost of building a nuclear power plant at a remote facility was greater than the cost of a diesel plant of equivalent capacity. As long as the savings of logistics' costs could be factored into the cost-effectiveness equation, military considerations would balance any additional expense entailed in using nuclear power. During the early 1960s, however, the savings in costs of transporting fuel was removed by OSD from the formula used to figure cost-effectiveness of nuclear plants vis-à-vis conventional power sources.

As a means of reducing cost inequity of nuclear power plants, during the 1960s, successive directors of the Army's Nuclear Power Program continued efforts to develop a standard plant design. In addition to lowering costs of development, procurement, and construction, such a plant would also minimize the retraining of crews for each individual power plant. But, while the research and development of a standard design progressed, the future of the Army's Nuclear Power Program began to be questioned for reasons unrelated to costs.

Although construction began on a 10,000-kilowatt floating nuclear power plant in January 1963, Camp Century was deactivated in July 1963. Despite his best efforts, the Chief of Engineers could find no military service willing to accept the plant in another location. Consequently, the Nuclear Power Field Office had to put the PM–2A's components into storage and ultimately submit the reactor vessel to destructive testing. Even more than the inability of the Army to re-site the Camp Century plant, the shutdown of the PM–1 at Sundance, Wyoming in April 1968 provided the most visible indication of the decline of the Army's Nuclear Program.

Originally, the Army had envisioned a large number of small reactors providing power to remote bases, particularly for the Air Force's early warning radar network. However, the development of intercontinental ballistics missiles during the 1950s and their deployment in the early 1960s made the DEW line and the need for small nuclear reactors to alleviate logistical problems obsolete. The Air Force's closing of the Sundance Radar Site, therefore, stands as a symbol of the decreased requirement for plants like the PM–2A.

By this time, the Army's Nuclear Power Program had suffered several research and development setbacks. The first occurred in January 1961, when the SL–1 experimental boiling water reactor experienced a nonnuclear explosion that destroyed the plant and caused the death of the three-man operating crew. As the first major accident, the tragedy marked the beginning of uncertainty for the Army's Program, as well as for the future use of the atom as a source of cheap, safe energy.

## NOTES

1. The discussion of the building and operation of the PM–2A came from the documentary and printed material in the PM–2A files in Box 6 of the ANPP records, unless otherwise cited. See also "Five on the Line."
2. Army Corps of Engineers, *News Bulletin #2*, Nov. 17, 1961 (cited hereafter as *News Bulletin*).
3. *News Bulletin #10*, August 7, 1962; *News Bulletin #14*, Feb. 5, 1963.
4. *News Bulletin #19*, Oct. 15, 1963.
5. Ibid.; *News Bulletin #28*, Dec. 13, 1963.
6. *News Bulletin #25*, July 31, 1964; *News Bulletin #27*, Sept. 15, 1964.
7. *News Bulletin #27*, Sept. 15, 1964.
8. *News Bulletin #28*, Oct. 15, 1964.
9. *News Bulletin #3*, Dec. 15, 1966.
10. The discussion of the building and operation of the PM–1 came from the

documentary and printed material in the PM–1 files in Box 6 of the ANPP records, unless otherwise cited. See also "Five on the Line."

11. *News Bulletin #5*, Feb. 15, 1962; *News Bulletin #6*, March 15, 1962; *News Bulletin #7*, April 13, 1962.

12. *News Bulletin #11*, Sept. 17, 1962; *News Bulletin #12*, Nov. 15, 1962.

13. *News Bulletin #17*, July 26, 1963.

14. The discussion of the building and operation of the PM–3A came from the documentary and printed material in the PM–3A files in Box 6 of the ANPP records, unless otherwise cited. See also "Five on the Line."

CHAPTER 6

# THE NUCLEAR PROGRAM IN DECLINE

The development of the SL–1 had its origins in the same Defense Department requirement for a small nuclear power plant that had produced the SM–1 and subsequent Army nuclear power plants. Colonel Lambert and his staff had selected the pressurized water reactor design for the Army's family of nuclear plants because they could be built within the state-of-the-art that existed in 1953. Moreover, the pressurized water reactor design had advanced more than the other system being developed and so was more readily adaptable to the Army's immediate requirements.

At the same time, within the feasibility analyses and conceptual designs at which Colonel Lambert had looked were two other systems that offered potential in terms of longer-range development. The first, the Gas-cooled Reactor system, ultimately became the focus of the Army's efforts to develop a nuclear power plant that would fit on a tractor trailer so that it could be hauled from place to place. The second, the Boiling Water Reactor System, was already the subject of intense development by the AEC itself, at the request of the utility industry, with principal research being done at the Argonne National Laboratories in Illinois.[1]

The Boiling Water Reactor offered special appeal to the military because of its interest in small, transportable nuclear power plants. The reactor's design eliminated the massive heat exchanger, a major component of the pressurized water reactor, which was heavy as well as expensive. Instead of using primary and secondary systems, the Boiling Water Reactor itself served as the steam generator providing steam directly to the turbine. Consequently, even though the Army's Nuclear Power Program put the preponderance of its efforts in the mid-1950s into the design and building of the SM–1 and subsequent pressurized water reactors, it also put some effort into bringing a Boiling Water Reactor into the complement of available types of nuclear power plants to meet specific military requirements.

Ultimately, the Reactor Development Division of the AEC assigned the Argonne National Laboratories the mission of preparing the design for a lower power reactor system based on the Boiling Water Reactor concept. While being only one-third to one-quarter the size of the SM–1, the plant was to be adaptable to the 300- to 1,000-kilowatt power requirements of a DEW line radar station. Because the mission was a development project, the AEC assumed the funding of the Reactor and in August 1955, the AEC's Chicago Operations Office selected Pioneer Services and Engineering Company of Chicago as the architect-engineer to design the plant's nonnuclear components. Concurrently, the Argonne National Laboratories designed and developed the reactor system. The power plant was to be sited at the AEC's National Reactor Testing Station, 8 miles east of Arco, Idaho.

Although the SL–1 was to be an experimental facility under the control of the AEC, the Nuclear Power Field Office received the assignment to operate and maintain the plant. Consequently, in the spring of 1956, the Office organized and trained the initial military crew. Since the Field Office expected that all three services would ultimately use this type of plant, it asked the Navy and Air Force to serve with Army personnel. The first crew included 12 Army, 3 Navy, and 4 Air Force men.

Like the first group of officers assigned the SM–1 when it went on-line in 1957, the officers who were to command the initial SL–1 crew took college nuclear engineering courses and attended the International School of Nuclear Science and Engineering at the Argonne Laboratories. They trained on the Argonne Research Reactor, which had started up in 1954, and on the Experimental Boiling Water Reactor, which had reached full power in December 1956. On their part, warrant officers and enlisted men took a special five-month course at the University of Virginia and then received on-the-job training at facilities of the Phillips Petroleum Company, which operated the National Reactor Testing Station.

In the meantime, site work for the SL–1 facility began in the fall of 1956. The general contractor, the Fegles Construction Company, began plant construction in 1957. Once their training had ended, the military operating crew joined the construction force to help install major pieces of equipment. From February 1 to April 25, 1958, the crew attended a plant orientation program and then assisted the Argonne Laboratory staff in checking instrumentation, electrical components, the steam system, and in conducting the initial critical tests. The SL–1 reached criticality on August 11, 1958 and began producing electricity on October 24.

The plant facilities consisted of the cylindrical reactor building, an adjoining support building, which contained the control room and auxiliary equipment, and several service buildings. Later on, a training building was added and the support building was enlarged. Most of the power plant equipment was located in the cylindrical steel building 38′6″ in diameter and 48′ high. Since the plant was expected to be air-transportable, no component measured more than 8′ × 9′ × 20′, nor weighed more than 20 tons.

The reactor building was built of steel plates, most of which were ¼-inch thick. The building did not function as a containment vessel in part because the site was in an unpopulated area. In addition, the building was part of the experiment since the SL–1 served as the prototype of an uncontained reactor that would be installed in remote military bases having a very low population density.

Carrying the experimental nature of the building one step further, it was set on dummy piles to simulate the type of construction that would be used in the Arctic in a permafrost area where the entire structure would have to sit on piles. The reactor vessel, fuel storage well, and demineralizer were located in the lower third of the building and shielded with gravel; it was anticipated that the on-site construction would use gravel as much as possible to cut down on logistical problems. The rest of the equipment and shielding for the reactor vessel occupied the upper two-thirds of the structure.

The SL–1 was the first nuclear power plant in the world to have aluminum-clad fuel elements. Previously, aluminum had been used only in the research and testing reactors, which had low heat levels because the metal had a low melting point. In the SL–1, however, a special aluminum alloy solved the problem. Subsequently, aluminum alloys became widely used in nuclear reactors. The core itself was designed to operate for three years without refueling, to be air-deliverable, and to be capable of being installed by hand. Unlike the Army's pressurized water reactors, the SL–1 had only five control rods.

Following the 500-hour full-power performance test that lasted until February 1959, the AEC awarded a contract to Combustion Engineering to operate the SL–1 and to conduct research and development studies aimed at improving the operating characteristics of the plant. Combustion Engineering in turn employed the military operating crew to continue running the plant. Thus, in essence, the SL–1 became part of the Army's Nuclear Power Program.

Once in full service, the SL–1 provided operational experience on a Boiling Water Reactor, developed plant performance characteristics, provided core burn-up data, trained military crews in plant maintenance, and tested components planned for use in subsequent reactors. In the fall of 1959, two groups from the Maritime Administration trained on the SL–1 in preparation for operating the first U.S. nuclear-powered merchant ship, the *Savannah*. By December 1960, when the plant had been in operation for more than two years, it had produced 931.5 megawatt days of thermal energy, which amounted to about 40 percent of the design life of the core.

On December 23, 1960, the plant crew shut down the SL–1 for routine maintenance, instrument calibration, installation of auxiliary system valves, minor plant modifications, and installation of flux wires in the core. Once the other work had been completed, installation of the flux wires began shortly after midnight on January 3. These wires were to be used to measure power distributions within the core as part of some investigations on reactor stability. The work was completed late in the day and the 4 P.M. to midnight shift was directed to reassemble the reactor in preparation for resuming operations the next morning.[2]

At 9:01 P.M., a violent upsurge of energy and a rapid formation of steam forced a column of water above the core and lifted the reactor vessel about 9 feet. The heat and nonnuclear explosion violently sheared all piping connections and completely destroyed the core. The subsequent investigation never was able to explain precisely what precipitated the accident. Two key factors stand out, however. For some time, the aluminum cladding on the control rods had been warped, causing the rods to stick on occasion. At the same time, the SL–1 had a unique feature. Unlike other reactors, which could not go critical with the removal of any one control rod, the SL–1 could go critical with the raising of only the central control rod.

From what could later be ascertained, one of the operators had begun to raise the central control rod the 1.5 inches needed to reconnect the drive mechanism. When the rod stuck, the man leaned over and tried to physically lift it the necessary distance. Instead, the rod apparently suddenly popped free. The operator, caught by surprise, could not control the rod. As a result, it rose out of the core with a surge, causing the reactor to explode from the virtually instantaneous superheating of the core water into steam.

The explosion triggered automatic heat-detection alarms, which sounded at 9:01 P.M., at the three fire stations of the National Reactor Test Station, and a radiation monitor alarm sounded in the nearby Gas-cooled Reactor Experiment (GCRE). When the fire department and the security forces entered the SL–1 fenced area, they found nothing visibly out of place. Upon entering the reactor support building, however, the assistant fire chief immediately detected radiation levels up to 25 roentgens per hour, but found no sign of the reactor crew. When the health physicist from the Materials Testing Reactor arrived, he entered the support building and proceeded through it to the reactor building where he found levels of 200 roentgens per hour at the foot of the stairway to the reactor. Because of the high levels of radiation, the decision to enter the reactor building to search for the plant personnel was not made until the Combustion Engineering plant health physicist arrived.

The health physicist and other personnel then entered the building in teams wearing protective coating. They found two of the crew, one of whom was still alive, on the floor near the reactor in a radiation field of approximately 1,000 roentgens per hour. Ignoring the radiation danger, rescuers pulled him out, but he died on the way to the hospital. Only then did a team of rescuers locate the third body, pinned to the ceiling of the reactor room by the central control rod. Working in tandum and remaining in the building for scant minutes, the rescuers brought out the first body during the night. However, it took several days to work out procedures for removing the impaled body.

Once the final body had been removed, the AEC, Combustion Engineering, and Army personnel began surveying the damage to the reactor and the extent of radiation. The reactor building had not been designed to serve as a containment vessel; however, it had functioned as such and no significant radiation was escaping beyond the immediate vicinity of the reactor building. It was also ascertained that the reactor was in a noncritical condition. Unfortunately, the physical condition of the reactor core, the location of the control rods, and the presence or absence of water in the pressure vessel were unknown. Therefore, no conclusion could be drawn as to whether or not the reactor might at some point sustain a renewed nuclear excursion.

From the end of the first phase of the recovery operation on January 10, 1961, to the end of the second phase in May, Combustion Engineering attempted to establish the nuclear status of the reactor core and confirm that no new nuclear excursion would occur. Given the very high levels of radiation found in the reactor building, most of the effort had to be done by remote control. Ultimately, it was determined that the pressure vessel contained no water and subsequent nuclear excursions could be prevented by keeping the vessel dry. On May 23, 1961 the AEC signed a contract with General Electric. At this time the company undertook to gather and evaluate data concerning the accident; remove the SL–1 core for examination, evaluation, and disposal; raze the reactor building completely; and decontaminate the SL–1 area to ready it for potential beneficial occupancy. These efforts lasted until July 27, 1962, when the area on which the SL–1 had stood contained no sign of a nuclear power plant or of radiation leakage.[3]

The contamination of the building and the consequent need to work by remote control accounted for the lengthy clean-up operation. To remove the pressure vessel, which was 16′6″ long and 4′6″ in diameter and contained the reactor core, workmen had to clear away many large pieces of equipment, remove the roof of the building, and cut out sections of the wall and floor. On November 29, 1961, using a large crane, they lifted the 13-ton piece of the plant 32 feet and swung it out of the building. As they did this, they pulled it inside a plastic bag to minimize the spread of contamination. Once out of the building, they placed the pressure vessel inside a cylindrical, 2-foot-thick concrete shielding tank mounted on a low-bed trailer. The next day, it was hauled 40 miles across the desert to a large hot cell (a heavily shielded enclosure where radioactive materials can be safely and remotely handled). Working with remote-control manipulators, technicians dismantled the radioactive vessel and core in an attempt to learn more about the accident. In the meantime, workmen demolished the reactor building and buried its contaminated components. The support buildings were decontaminated by vacuuming and scrubbing.[4]

In investigating the cause of the accident, the AEC had initially appointed a board of inquiry, which ultimately interrogated more than 475 people. Neither the Agency's Report of May 1961, nor the more detailed General Electric Report of July 27, 1962, were able to give a definitive explanation for the SL–1 accident. However, the letter from the SL–1 Board of Investigation of May 10, 1961, transmitting its Report to the AEC's General Manager, pinpointed the condition of the reactor itself as creating an environment conducive to disaster. According to Curtis Nelson, the Board's Chairman, the SL–1 "reactor core and the reactor control system had deteriorated to such an extent that a prudent operator would not have allowed operation of the reactor to continue without a thorough analysis and review, and subsequent appropriate corrective action, with respect to the possible consequences or hazards resulting from the known deficiencies." The Board believed that such review and actions should have resulted "in modifications to design, administration, and operation sufficient to insure that there was no potential hazard greater than contemplated in the original hazards report and review, before reactor operation was resumed."[5]

While the Board did not rule out a nonnuclear event as the cause of the accident, it concluded that a nuclear excursion appeared a "more credible" cause. Whether or not the excursion was in fact initiated by an operator trying to withdraw the central control rod as seemed most plausible, the Board stated that this "hypothesis is useful in discussing the relationship among the various factors which would have, but may not have, contributed to the accident." In the prime scenario, the Board observed that the reactivity of the reactor had increased due to a loss of boron from the control rods. As a result, the distance the central rod had to travel to produce an excursion had decreased to the degree that the reactor could go critical more quickly than it would have done if there had been a larger shutdown margin of reactivity due to less mechanical loss of boron. In the event that when the control rod suddenly became unstuck due to the operator's efforts its unexpected upward motion exceeded the actual distance, if not the theoretical distance, needed to cause the nuclear excursion.[6]

Both the Board's Report and the final General Electric Report suggested a series of measures that reactor designers, builders, and operators should take in order to avoid similar accidents. First, reactor designers should avoid reactors that could become critical when only one control rod is withdrawn. In addition, if a then-operating nuclear reactor offers the potential for going critical in the manner of the SL–1, operators should incorporate precautions into the control procedures. In any case, if a reactor is experiencing control or other operating difficulties as the SL–1 had been having—such as a change in reactivity and sticking control rods—it should be shut down until effective remedial steps could be taken.[7]

Noting the divided control of the SL–1 between the AEC and the Army because of the experimental nature of the reactor, the two reports stated that any decision to continue operation of the reactor should have been made at headquarters level, not by personnel at the reactor plant. In the future, detailed procedures for conducting reactor maintenance should be developed and performed under the constant supervision of a technically trained person. In addition, an operator should always remain in the control room when work is being done on the core, control rods, or control-rod drive mechanism. At the time of the SL–1 accident, all three crew members were in the reactor room. In addition, every reactor should have a periodic safety review by an independent group and responsibility for health and safety should be clearly defined.[8]

In addition to the conclusions reached by the Board, the AEC evaluated the cost of continuing the Boiling Water Reactor Program against the benefits to be obtained from continuing the development of the design concept. When the AEC determined that it was not worthwhile to continue funding the project in light of budgetary demands, it phased out the program and all studies were placed on the shelf for possible future use. According to Gen. Gribble, who was head of the Army's Nuclear Program during these considerations, the Boiling Water Reactor concept remained "a very attractive system" whose development had progressed "quite well." Following the lead of the AEC, the Army nevertheless stopped funding its development program for the design.[9]

In fact, the Army had several other development projects competing for its research dollars. Each in its own way represented an esoteric attempt to use nuclear energy to supply military needs. At the height of the Cold War, however, appropriations remained relatively easy to obtain, especially for projects offering private as well as military applications. As a result, the Nuclear Program was able to explore the possibilities of several types of reactors for a variety of purposes.

The effort to develop a Military Compact Reactor (MCR) grew out of the Army Transportation Corps' interest in using nuclear energy to power heavy overland cargo haulers in remote areas. Initial studies begun by the Nuclear Development Corporation began for the Transportation Corps in December 1955, and focused on various reactor and vehicle concepts, including a small liquid-metal-cooled reactor.[10] Following up on this research, the Army Reactors Branch of the AEC awarded its own contract to General Motors in March 1956 to study the application of nuclear power to the overland train. Two years later, the AEC awarded three additional contracts for studies to select a reactor design combining compactness and the lightest weight suitable for the land train. All three contractors recommended liquid-metal-cooled reactors coupled with open-cycle air turbines. By mid-1958, however, the studies concluded that the weight of the reactor and its shielding would make the power plant unsuitable for the proposed cargo hauler.

If the Military Compact Reactor, as the project became designated, proved impractical to power a transportation vehicle, the military still saw the idea of a small nuclear reactor as the source of mobile electric power for field forces. Consequently, from 1958 to 1963, the Army and the AEC pursued the development of a reactor that could be readily moved and would produce 2 to 3 kilowatts of electricity. By April 1963, however, the AEC and the Department of Defense had combined to spend just over $8 million on the project while still in the design stage. At that point, it was estimated that the total cost of development would run to $77,700,000 with 80 percent coming from the AEC. In the face of this reality, the AEC announced it would only continue development of the concept at a much lower level, citing the need for more specific military requirements to justify great expenditures and indicating its doubts that the proposed reactor coolant could be obtained economically.

From this time on, the Military Compact Reactor (MCR) program became tied to the development of the Nuclear Power Energy Depot, which was then being considered as a means of producing synthetic fuels in a combat zone for use in military vehicles. In November 1963, an Army Staff study submitted to the Department of Defense concluded that the most promising means of employing MCRs was as the power source for the Energy Depots. As a result, the MCR found itself bound to a program that offered an uncertain feasibility.[11]

Essentially, the Energy Depot was an attempt to solve the logistical problem of supplying fuel to military vehicles on the battlefield. While nuclear power could not directly supply energy to small, widely dispersed units, planners proposed to use nuclear reactors to provide power to manufacture, under field conditions, a synthetic fuel as a substitute for conventional hydrocarbon fuels. The nuclear plant would be combined with a fuel production system to turn readily available elements such as hydrogen or nitrogen into fuel that could be used as a substitute for gas or diesel fuel in cars, trucks, and other vehicles.

Of the fuels that could be produced from air and water, ammonia and hydrogen seemed to offer the best possibilities as substitutes for petroleum. By electrolysis, water could be broken down into hydrogen and oxygen and then the hydrogen could be used to produce energy. As an alternative, the nuclear plant could supply power to produce nitrogen through the liquefaction and fractional distillation of air. The nitrogen could then be combined with hydrogen to form ammonia as fuel for internal-combustion engines. Finally, although thought was given to using Military Compact Reactors to generate electricity to charge batteries for electric-powered vehicles, that option was contingent on the development of suitable battery technology.

The Corps of Engineers and the Army Materiel Command (AMC) shared the responsibility for developing the Nuclear Powered Energy Depot concept. The Engineers were to develop the power plants and fuel-production machinery and the AMC was to develop the vehicle engines and related equipment. In fact, the Engineers investigated the feasibility of using ammonia as a motor fuel. Between January and July 1964, SFC Herbert Kappel and SFC Frederick Urbin, both of the Academic Training Branch of the Nuclear Power Field Office, undertook to modify gasoline-powered engines to run on ammonia. Ultimately, they converted the engine of a half-ton truck to use ammonia. On July 27, they road-tested the vehicle at speeds up to 25 miles-per-hour and developed enough power to climb an 8-percent grade. Because ammonia has a lower energy output than gasoline, the two experimenters also used acetylene to enrich the ammonia and improve the engine performance.

The success of the experiment notwithstanding, the feasibility of the Nuclear Powered Energy Depot itself remained in doubt. Aerojet-General Nucleonics found shortcomings in the proposed method of producing ammonia from water and air. During 1965 and 1966, the Stanford Research Institute studied the idea of using electrical energy from a nuclear power plant to recharge electric cells for electric-powered military vehicles. Ultimately, however, the engineers themselves exposed the problems with the Energy Depot concept. The Engineer Strategic Studies Group within the Office of the Chief of Engineers completed a study on the cost-effectiveness of the Energy Depot within the current or projected 10-year state of technology. The Group concluded that although feasible, the Nuclear Powered Energy Depot systems would require equipment that probably would not be available during the next decade. As a result of these findings, the Assistant Secretary of the Army for Research and Development ordered the suspension of further development of the Energy Depot until it became economically attractive and technologically possible.[12]

From the beginning of the Transportation Corps' interest in a small, compact reactor in 1955, to the end of the efforts to find a use for such a plant, the MCR never went beyond the design stage. In contrast, the ML–1 project ultimately became operational. Apart from the design, construction, and operation of the six plants that formed the heart of the Army's Nuclear Program, the effort to develop a nuclear plant combining the logistical merits of a nuclear reactor with the mobility of a diesel generator probably occupied more of the Nuclear Power Field Office's time and interest than any other project during the height of activities.

The search for the ideal plant began in 1956 with the start of the ML–1 research and development program. The goal was to build a mobile, low-power nuclear plant that could accompany troops from place to place and provide power to command and communications centers, evacuation hospitals, depots, and radar and weapons systems. Initial plans called for a plant with a capacity of several hundred kilowatts that would be shock-mounted and transported on low-bed trailers. The ML–1 was expected to weigh about 40 tons and replace conventional (diesel) generating equipment that weighed only 20 tons but consumed 4 tons of fuel oil per day. A single supply of nuclear fuel would eliminate the logistical effort required to supply 400,000 gallons (1,600 tons) of fuel oil. In addition to the Army's special interest in the reactor, the AEC supported its development because it would contribute to the technology of both military and small commercial power plants.

To develop a nuclear plant small enough for full mobility required a whole new technology. Early studies suggested that a gas-cooled reactor combined with a gas-turbine power-conversion system would be far lighter than a pressurized water reactor and would eliminate the need for large quantities of water to cool it. Because the gas would become radioactively contaminated and thus unsafe to be discharged into the atmosphere, it would be necessary to develop a closed-cycle turbine. No such turbine had been built or operated in the United States, however, and those built in Switzerland offered little applicable experience because they were used in commercial power stations. Under the circumstances, the Nuclear Power Program had to begin by testing the concept of the ML–1. A demonstration model of the reactor and power-conversion system had to be designed, fabricated, and tested.

Under the policy already established, the AEC assumed the responsibility for developing and financing the gas-cooled reactor for the ML–1 and the Corps of Engineers took the job of designing and building the power-conversion system. Given the lack of a working example, both agencies could not start with a prototype unit as they had done with the SM–1. Instead, they had to assemble technical information through a series of experiments.

As its first step in obtaining a reactor, the AEC awarded a contract to design, fabricate, and operate a Gas-cooled Reactor Experiment (GCRE) at the National Reactor Testing Station to Aerojet-General Corporation, which had been the lowest of 20 bidders. Once the company had completed the project in 1959, the AEC used it to test reactor behavior under various conditions, evaluate reactor components, test fuel elements, and obtain technical information about the reactor design.

On its part, the Engineer Reactor Group had to start from scratch to develop the turbine-compressor unit and various heat-exchange units that would ultimately be coupled with the reactor. In June 1957, the Engineers awarded a contract to Aerojet-General to construct a Gas Turbine Test Facility at Fort Belvoir. The contract called for the design, construction, procurement, installation of components (except for the turbine-compressor set furnished by the Government), and the acceptance-testing of the system. When the facility was completed in 1959, the Nuclear Power Field Office used it for developmental testing of closed-cycle gas turbines and related equipment.

The Gas Turbine Test Facility occupied an entire building and although it did not ever remotely resemble the completed turbine generator that would be hooked up with the ML–1 nuclear reactor, it contained all the components that would ultimately make up the turbine segment of the power plant. The components included a closed-cycle turbine-compressor set, recuperator, precooler, generator, and oil-fired heater with control devices to simulate the way a reactor could supply heat to the completed prototype power plant.

Testing began with experiments that would help select the coolant gases and materials for the turbine that could withstand corrosion and extremes of temperature. Ultimately, nitrogen was chosen because it is relatively inert and easily obtained. Moreover, because air is 78 percent nitrogen, engineers expected to be able to modify the turbine at some point so that it could operate on air. Other tests at the turbine facility at Fort Belvoir and the reactor test facility in Idaho enabled Aerojet-General to acquire the data needed to develop a design for the ML–1 prototype. The company built the reactor at its Downey, California plant and trucked it on an Army low-bed trailer to the National Reactor Testing Station where it attained criticality on March 30, 1961.

Following 18 months of testing, during which time the reactor was connected to the completed gas turbine and generator, the ML–1 operated as a power plant for the first time on September 21, 1962, albeit producing only a few kilowatts of electricity. In doing so, however, the ML–1 stood as the smallest nuclear power plant on record to produce electricity. Its reactor operated at the highest temperature of any nuclear installation, heating its gas to a temperature of 1,225°F. It was also the first nuclear power plant on which the reactor was coupled directly to a closed-cycle, gas-driven turbogenerator.

The ML–1 was designed to produce 300 to 500 kilowatts and to operate for 10,000 hours between fuel changes and 50,000 hours between overhauls. The Army expected to mount ML–1–type power plants on six skids and transport them as an integrated package by military semi-trailers, railroad flatcars, or barge. If the plant were split up, it could be moved by transport aircraft. The plant was designed to be set up and ready for operation within 12 hours after its arrival or prepared for removal within 6 hours. Unfortunately for the Army, the contractor, and the Pentagon, the ML–1 prototype suffered from regular technical problems that continued during its initial assembly and test period, and then during its experimental testing phase.

The results of the first 100-hour high reactor power operation test illustrate the problems that continued to appear within the plant's systems. The ML–1 had reached full power for the first time on February 28, 1963. Following its shutdown on March 4, the crew discovered that coolant gas was leaking into the moderator water. Subsequent investigation revealed the leak was probably due to a mechanical failure of a seal weld in a structure member beneath the pressure vessel. As a result, the reactor had to be defueled and disassembled in order to reach and repair the problem. Consequently, the ML–1 did not begin generating power again until the spring of 1964.[13]

Once back in operation in April 1964, the ML–1 underwent a series of rigorous tests that lasted until January 1965. Full-power and limited-endurance tests took place from mid-April to the end of May 1965. During this time, it operated only 63 percent of the total elapsed time of 1,054 hours, after which it was again shut down for repairs. It then underwent the thermodynamic performance test, beginning on September 1 and ending on October 20, when the plant was shut down for scheduled maintenance. Testing resumed three days later and continued to the end of November. During the next endurance test from December 6 to 17, the plant performed well, but experienced two unscheduled shutdowns due to cold weather. The next test, a full-power and after-heat evaluation test, continued from December 17, 1964 to January 26, 1965.[14]

Meanwhile, the Gas Turbine Test Facility at Fort Belvoir proved insufficiently versatile for all the required testing and experiments with the components of the ML–1. Consequently, the Engineers redesignated and rebuilt the facility, redesignating it the Advance Power Conversion Experimental Facility. It included a turbine-generator, pre-cooler, recuperator, heater (to simulate the reactor), electrical generator, emergency power equipment, and associated instrumentation. Within the building, these components were spread out as far as possible to provide the most favorable conditions for studying their characteristics. Once the facility began operating on February 16, 1965, the Nuclear Power Field Office was able to obtain valuable results in regard to the operation of closed-cycle gas turbine systems.

At the same time, however, despite almost three years of operation by the summer of 1965, the ML–1 continued to suffer from regular mechanical breakdowns. In addition, during its test life, the plant proved to emit radioactivity that would make its use in the field impractical, if not dangerous, to personnel. In light of the ongoing difficulty in eliminating problems and demonstrating the plant's reliability and suitability for military use in the field, the AEC initiated a review of the entire ML–1 program.

On its part, the Agency concluded that the probability of achieving program objectives in a timely manner and at a reasonable cost was not high enough to justify continued funding of its portion of the project. On the military's side, cutbacks in funds for long-range research and development because of Pentagon commitments to the Vietnam War suggested to the

AEC that the ML–1 could not continue to receive Defense Department funding. As a result, the AEC phased out its part of the program as of FY 1966. Following additional turbine testing, the Nuclear Power Field Office also ended its portion of the project on June 30, 1966.[15]

With the phasing out of the ML–1 program, the Nuclear Power Field Office ended its efforts to develop small, stationary, and mobile reactors that could be used as a source of power in field operations. From its experiences with the SL–1, MCR, and ML–1 programs and from its efforts to develop a nuclear plant for NASA's Lunar Landing Program, the Engineers learned that the development of complex, compact nuclear plants of advanced design was expensive and time consuming. They found that the costs of developing and producing such plants are in fact so high that they can be justified only if the reactor has a unique capability and fills a clearly defined objective backed by the Department of Defense. In addition to having an interest, however, the Army and the Pentagon had to be prepared to furnish financial support commensurate with the AEC's development effort on the nuclear side. Nevertheless, despite these set-backs, in the mid–1960s the Nuclear Power Program still seemed viable, especially as work on the first floating power plant, the *Sturgis*, neared completion.

## NOTES

1. The information on the building and operation of the SL–1 comes from the SL–1 files in Box 6 of the ANPP records, particularly the following: U.S. Atomic Energy Commission, "Report on the SL–1 Incident, January 3, 1961," n.d. [May 1961]; Frank K. Pittman, "Remarks Prepared for Presentation to Representatives of Industry—Review of Available Information on the SL–1 Incident," January 24, 1961; *Final Report of the SL–1 Recovery Operation,* July 27, 1962; Interview with Col. James Lambert, February 16, 1963, Box 6.

2. Ibid. The information on the SL–1 incident comes from the same sources.

3. *Final Report of the SL–1 Recovery Operation.*

4. Ibid.

5. Chairman, Curtis A. Nelson, SL–1 Board of Investigation, to AEC General Manager, A. R. Luedecke, May 10, 1961.

6. Ibid.

7. Ibid.; *Report on the SL–1 Incident; Final Report.*

8. Ibid.; *News Bulletin #12,* Nov. 15, 1962. This section was also based on interviews with Bill Soward of FESA who worked on the SL–1 during the shift prior to the accident and Alan Van Den Berg, a civilian veteran of the nuclear program. Author interview with Bill Soward, December 1979, March 1980, July 1980, Box 8; author interview with Alan Van Den Berg, March 1980, June 1980, Box 8.

9. Interview with Brig. Gen. William C. Gribble, Jan. 23, 1963.

10. The discussion of the Military Compact Reactor came from the MCR file in Box 5, ANPP.

11. The discussion of the Nuclear Power Energy Depot came from the NPED file in Box 5, ANPP, as well as *News Bulletins* cited below.

12. *News Bulletin #9,* Dec. 27, 1965; *News Bulletin #3,* Dec. 15, 1966; Director, USAERG, Report to the Chief of Engineers, May 15, 1967.

13. *News Bulletin #15,* April 25, 1963; *News Bulletin #19,* Oct. 15, 1963; *News Bulletin #21,* Jan. 15, 1964; *News Bulletin #25,* July 31, 1964.

14. *News Bulletin #27,* Sept. 15, 1964; *News Bulletin #28,* Oct. 15, 1964; *News Bulletin #29,* Nov. 15, 1964; *News Bulletin #30,* Dec. 15, 1964; *News Bulletin, #1,* vol. 2, Jan. 15, 1965; *News Bulletin, #2,* Feb. 15, 1965.

15. *News Bulletin #8,* vol. 2, Nov. 5, 1965; *News Bulletin #1,* vol. 3, May 22, 1966; *News Bulletin #2,* Sept. 1966.

## Chapter 7

# Transition

The concept of mounting a nuclear power plant on a barge had its origins with the inception of the Army's Nuclear Program. In advocating the early construction of a floating nuclear facility, Gen. Nichols had the example of the military's use of barge-mounted power plants during World War II and their later deployment in Europe, the Philippines, and Korea. In fact, the utilization of a floating source of electricity actually predated World War II.[1]

In 1929, the New Hampshire Service Company had converted the 5,000-ton freighter the *Jacona* into an oil-fired 20,000-kilowatt power plant. During World War II, the Navy had operated the *Jacona* at Pearl Harbor. The Army later took custody of the power plant for use in Korea and then at Okinawa. The wartime demand for emergency sources of power led to the construction of four powerbarges during 1942 and 1943, each supplying 30,000 kilowatts. Initially, all four operated in the Ohio and Mississippi valleys and in the Great Lakes, but ultimately each served abroad, with two seeing duty in Belgium after the Normandy landings.[2]

The operating barges provided electric power where land-based construction would prove difficult, expensive, or would not provide power in time to meet short-term needs during the reconstruction of war-destroyed permanent power facilities. With these needs in mind and World War II operations as a model, the Corps of Engineers requested the Marine Design Division in the Philadelphia District to design a barge-mounted conventional power plant in 1952. By the time the assignment was completed, the Army's Nuclear Program was underway and the planners noted that an atomic reactor could be substituted for the conventional power source. In fact, a conventional barge-mounted power plant would suffer from the same shortcomings, in regard to expense and logistics problems of delivering fuel, as land-based, electric-generating plants in remote areas. In any event, nothing came of the suggestion and the plans remained on the shelf for the next several years.[3]

However, as the Army's Nuclear Program developed and the SM–1 began operations, the idea of putting a nuclear power plant on a floating facility reemerged as a particularly attractive idea. Ultimately, during 1959, the Marine Design Division dusted off its blueprints for a powerbarge and, with help from a Nuclear Power Field Office engineer, began preparing specifications for a floating power plant. This plant called for a 10,000-kilowatt pressurized water reactor to be mounted on the hull of a World War II Liberty ship. The project soon suffered a delay due to insufficient funds, but in April 1960, the Army's Research and Development Office provided the OCE $600,000 to complete the design for the nuclear plant. Later in the year, seven companies submitted bids for the facility. On April 3, 1961, the Martin Company (later Martin-Marietta Corporation) received the contract for the design phase, with the option to undertake subsequent portions of the procurement and construction of the power plant.[4]

The contract, based on a negotiated low bid of $16,998,127, was to be administered in eight phases: design, construction of the power plant, procurement of the nuclear core, test operation, repair parts and tools, training, technical manuals, and a documentary film. The Army's design specifications, on which the bids were made, called for a plant that would perform a variety of missions, including beachhead and invasion support, supply of power for civil defense or civil affairs, and support of missile tracking or testing facilities. In the contract, the Government retained the option of deleting any or all phases of the contract following the initial design stage. On its part, Martin could exercise its options for subsequent stages of the project once it completed the design phase, scheduled for August 1962.[5]

According to the Army's original plans, the reactor, consisting of 32 fuel elements, would be placed on a revamped Liberty ship hull chosen from the mothball fleet in the James River across from Fort Eustis, Virginia. The hull was to be cut and 213 feet of the midsection were to be removed and scrapped. By removing the engines and turning the ship into a barge that required towing, the Army hoped to avoid any claims that Adm. Rickover might lay to the power plant on the grounds that the vessel belonged in his nuclear Navy realm. In any case, a new midsection containing the power plant was to be fabricated and joined to the original bow and stern, producing a craft of 406 feet.[6]

During the design period, the Martin Company benefited in its work from the Nuclear Power Field Office's experience operating the SM–1. Whereas the company's original proposal remained essentially intact, changes were made in both the arrangement of machinery and refueling procedures. Of particular importance was their decision to incorporate all "Hot" areas under a single control and decontamination point and to place all refueling operations under decks. In addition, in working on the safety aspects of the facility, Martin worked with the AEC's Division of Licensing and Regulation, which was paying attention to the MH–1A because of its groundbreaking nature. On August 23, 1962, the Commission's Advisory Committee on Reactor Safeguards received a briefing on the development of the power plant and later advised the AEC's Chairman, Glenn Seaborg, that the MH–1A could safely be tested at Fort Belvoir.[7]

With the design completed and approved, the Martin Company exercised its option for the construction, core fabrication, testing, and the making of a documentary film recording phases of the project. In turn, on January 18, 1963, the Philadelphia District Engineer issued notice for the contractor to proceed with the construction of the MH–1A. The cost of these phases of the contract totaled $15,800,000. Delivery of the plant to Fort Belvoir was scheduled for the summer of 1965, where it would undergo six months of test operations before its first siting at an as-yet-undetermined locale.[8]

Once the Martin Company received the go ahead, it awarded subcontracts to several other firms; the most important went to the Alabama Drydock and Shipbuilding Company of Mobile, Alabama which was to build the *Sturgis.* Because the barge would be constructed in Alabama, in March 1963, the Army selected the Liberty ship, the *Charles H. Cugle*, from the Mobile Reserve Fleet for conversion instead of picking one from the James River Fleet as originally planned. On April 15, the *Cugle* was towed to the subcontractor's shipyard and work stripping down the hull began almost immediately. Construction of the new mid-body that would contain the power plant began in July with completion of the Sturgis still scheduled for the summer of 1965.[9]

By June 1964, the containment vessel that would house the primary reactor system was ready for installation within the new hull. The spheriod cylinder weighed 350 tons and had been fabricated from 1-inch-thick steel plates reinforced with curved I-beams at the Delta Southern Company at Baton Rouge, Louisiana. The only crane on the southeastern seaboard capable of lifting such a load was located in the Avondale Shipyard at New Orleans. Consequently, both the containment vessel and the new hull had to be shipped by water to New Orleans.

The new midsection could simply be towed from the shipyard in Mobile. However, the containment vessel had to first be moved from the fabrication plant to the river bank, which proved a considerable problem not only because of its weight, but also because it was 43′ long and 31′ in diameter. During the nearly mile and a quarter trip first on a four-lane highway and then cross-country for an eighth of a mile on June 16, more than 120 overhead electric and telephone wires had to be removed and then replaced. The vessel itself was supported by a cradle strung between two specially rigged tractor-trailor trucks. The main load rested on 4 dollies with a total of 96 wheels and the tractors each had 10 additional wheels. Between the overland journey and the move down the Mississippi on July 1, the containment vessel received a paint job inside and out with spray resin paints. The "big lift" itself took place on July 3 at Avondale. Following the completion of welding the next day, the hull was returned to the Mobile shipyard to continue its construction.[10]

By the July 24 Progress Review Meeting at the Martin Company's Nuclear Division in Baltimore, however, the contractor acknowledged that it had fallen about 2 percent behind its self-imposed construction schedule. Nevertheless, the Company detailed its plans for recovering lost time and discussed its recently completed design for the MH–1A's reactor core. Fabrication of fuel elements and control rods began in September and by the middle of the month construction had passed the half-way mark. Despite this, at the September 16 Progress Review Meeting at the shipyard in Mobile, Martin said it would have difficulty meeting the July 1965 delivery date required by the contract.[11]

In November, construction of the MH–1A passed the 60 percent completion mark with work continuing on the barge in Alabama and with the reactor fabrication in the Martin plant in Baltimore. The same month, the Chief of Engineers Office announced that the MH–1A would be named for the late Lt. Gen. Samuel D. Sturgis, Jr., who had died on July 5, 1964. Gen. Sturgis had been Chief of Engineers from 1953 to 1956, during which time the Army had begun its nuclear power program.[12]

In addition to the continuing construction of the *Sturgis,* work got underway at Fort Belvoir to prepare the docking site for the barge during its test period. In early November, the Norfolk District Engineer let a contract to the American Dredging Company of Philadelphia to prepare a channel and base for the barge on Gunston Cove, just below the site of the SM–1. The Company moved the *American,* the world's most powerful hydraulic dredge, to Fort Belvoir later in the month and began work on November 30. And, as part of the same preparation, in November the District Engineer also awarded a contract to the Harwell Construction Company of Orange, Virginia, to construct the power line from the *Sturgis* to the Virginia Electric Power Company (VEPCO) power lines and to build the needed pier and access road.[13]

Apart from the proximity to the testing procedures the site offered several advantages for the Nuclear Power Field Office once the *Sturgis* arrived. Gunston Cove was within the restricted area that enclosed the SM–1 and normal security forces at Fort Belvoir would be able to keep unauthorized persons away from the powerbarge. After it went into operation during the test period, the *Sturgis* would be able to supply power for Fort Belvoir, saving it some cost of buying electricity from VEPCO. On the other hand, with both the SM–1 and the *Sturgis* producing electricity, some surplus power would be available and could be disposed of through the power grid to VEPCO at a rate equal to its lowest production cost.

By mid-December, however, reports from the contractor indicated that the construction schedule had now slipped by three months and the Martin Company now expected to deliver the *Sturgis* to Fort Belvoir in early October 1965. The Company attributed the delay to complications arising in procurement of various conventional components of the plant. Nevertheless, the Nuclear Power Field Office continued its preparations for training *Sturgis's* crew and for taking over operations of the plant following completion of its testing. As an initial step, in the spring of 1964, the Army Engineer Research and Development Laboratories (ERDL) at Fort Belvoir began the procurement process of an MH–1A training simulator designed to provide control room operator trainees with the " feel" of actual operation of the MH–1A's power plant.[14]

The Martin Company, as the prime contractor, prepared the design specifications for the simulator that was to contain an electronic computer and control consoles for both a student operator and the instructor. ERDL used widespread publicity in an effort to obtain maximum competition among qualified suppliers for building the simulator. Ultimately, ERDL sent the design requirements to 49 companies and received 8 proposals it deemed acceptable. The 8 successful companies were then issued an invitation on June 9, 1964 to bid on the contract. On June 24, the ERDL selected as the lower bidder the General Atomic Division of General Dynamics Corporation. The $307,000 contract for building the simulator included the provision for supplying drawings, manuals, training, and repair parts for one year of operation.[15]

By February 1965, dredging was completed at Fort Belvoir for the *Sturgis's* berth and work was continuing on the power line and pier construction. However, construction on the *Sturgis* was only 70 percent complete. In June, construction had reached approximately 79 percent, with nearly all major items of equipment then on board the barge. The remaining work consisted of installing the pipe lines and electrical connections. According to the Martin Company the schedule was no longer slipping and it informed the Nuclear Power Field Office that the *Sturgis* would arrive at Fort Belvoir in late December 1965 or early January of the next year.[16]

In anticipation of its arrival, the Nuclear Power Field Office reorganized the command structure for the *Sturgis*. On October 15, 1965, the MH–1A Floating Nuclear Power Plant Branch, formerly a part of the Field Office's Operations Department, became a separate element of the Nuclear Power Field Office. The new Branch operated like the SM–1 Branch with the same general authority and functions. Capt. Robert Bunker was appointed Officer-in-Charge of the Branch.[17]

Ultimately, because of continued delays in completing the *Sturgis* at the Alabama Drydock and Shipbuilding Company, the Nuclear Power Field Office decided the barge should be brought to Fort Belvoir where final work could be done under supervision of the Martin Company and the Engineers.

As a result, a settlement was reached with the subcontractor and the *Sturgis* was towed from Alabama in early April 1966; it arrived at Gunston Cove in the early hours of April 22, during a heavy rainfall.[18]

By then, the simulator had arrived and was in the final stages of installation. In addition to providing a detailed simulation of the design performance of the MH–1A, the simulator's computer had the capability of being operated to simulate other nuclear plants or problems that might arise in the Army's Nuclear Program. General Atomics Division completed work on the simulator in August 1965 and the Nuclear Power Field Office accepted the equipment. Final technical manuals remained in preparation, with delivery expected in early fall.[19]

Even after the *Sturgis* arrived at Fort Belvoir and the Army could work with Martin-Marietta (the Martin Company's new name), problems continued to plague the project. Even though the contractor completed all construction work during the summer, the date for the reactor's going critical slipped from September 1966 to January 1967. The delay resulted from significant problems with the control rod drive mechanisms, which suffered damage on two occasions during testing. Finally, on January 24, 1967, the MH–1A went critical and testing began.[20]

Once the reactor reached full power, however, tests revealed a major flaw in the shielding that permitted neutron radiation in the control room to rise substantially above expected limits. Investigation determined that the radiation was scattering on the lead shield under the reactor and Capt. Bunker ordered the reactor shut down until additional shielding was designed and installed. With this done, the *Sturgis* began the testing period intended to demonstrate its reliability, safety, and capability to provide a large block of electric power to support military operations. The testing lasted for five months, during which time the MH–1A operated at levels up to its full-rated power of 10,000 kilowatts. Finally, on July 25, 1967, the Army accepted the floating power plant from the Martin-Marietta Company.[21]

For the next 11 months, the *Sturgis* remained at Fort Belvoir generating power for the base while the Corps of Engineers sought a permanent suitable site. As part of the siting process, on March 15, 1968, the Chief of Engineers activated the 535th Engineer Detachment (Floating Power Plant, Nuclear) with an authorized strength of 68 men. Two-thirds of the crew were skilled nuclear power specialists and graduates of the Army Power Plant Operator's Course at Fort Belvoir, who had been assigned to the *Sturgis* since its arrival from Alabama. The balance of the contingent made up the deck force and various administrative and supply personnel.[22]

By the spring, the State Department and the AEC had entered into negotiations with the Panama Canal Company for use of the *Sturgis* in the Canal Zone. In 1968, several hydroelectric and conventional power plants with a combined capacity of some 100 megawatts provided electricity to the Canal Zone. However, insufficient reserve capacity existed to permit the largest generator unit to shut down for maintenance or emergency repairs without interrupting power to military and civilian customers. In addition, the Gatun hydroelectric facility competed directly with the Canal locks for water needed to operate them. Two factors compounded the problem. Due to the closing of the Suez Canal and the increased shipping from the U.S. East Coast to Vietnam, traffic on the Canal had recently increased considerably. This put additional demands on Gatun Lake to supply water for the locks. Consequently, during the dry season from December to May, the level of Gatun Lake had fallen drastically, which in turn curtailed operation at the Gatun hydroelectric station.[23]

To meet the demands for additional power, the Panama Canal Company had to purchase electricity from a State of Panama power company. The use of the *Sturgis* offered the Canal Zone the chance to cut back on these purchases. Ultimately, the Army agreed to lend the *Sturgis* to the Panama Canal Company, but only after the State of Panama was assured that the nuclear reactor posed no threat to the surrounding area. According to AEC regulations, all solid wastes and spent fuel were to be returned to the United States for disposal and the Canal Company agreed to pay for the electricity produced.[24]

The *Sturgis* left Fort Belvoir under tow at the end of July and arrived on August 7, 1968 at Gatun Lake, where it was moored adjacent to the hydroelectric station. The crew required two months to unwrap the power plant from its travel mode and check out components before it went on the line on October 5, 1968. Connected to the Panama Canal Company power grid by underwater cables, an unknowing person would have had no idea that the Sturgis was a nuclear power-generating facility.

To further alleviate the power shortage in the Canal Zone, the Army assigned the powership *Weber* to the 535th Engineer Detachment. The *Weber* had been converted from its previous role as a Navy Advance Base Drydock to a 20,000-kilowatt floating power plant between 1966 and 1968. It was then assigned to Kwajalein atoll in the Marshall Islands to supply electricity on the island. In July, while on the way, the Army diverted the *Weber* to the Panama Canal Zone where it was moored in Balboa on the Pacific side of the waterway. On May 15, 1969, the Army permanently assigned the *Weber* to the Corps of Engineers; the Chief of Engineers subsequently assigned the plant to the Army Engineer Reactor Group.[25]

Perhaps as much as anything, the *Weber's* assignment to the Nuclear Power Program marked the end of the Army's comittment to the atom as a source of power for military needs. The *Sturgis* had achieved Gen. Nichols' original aim of developing a mobile source of power capable of providing electricity to areas where other sources were not readily available. Nevertheless, whereas the *Sturgis* was to demonstrate its reliability for the next eight years, it became the last nuclear plant built by the Army. Moreover, by the time the MH–1A left Fort Belvoir, the Nuclear Program had begun to wind down even if its future demise was not yet readily apparent.

The PM–1 ceased operations on April 11, 1968 when the Air Force shut down the Sundance Radar facility and the SM–1A came close to being decommissioned because of the embrittlement of its pressure vessel from radiation. Initially, the Fort Greely facility was scheduled for shutdown because of the problem. The Army's research and development program discovered, however, that the embrittlement of the carbon steel of the reactor vessel could be removed by annealing the steel. In August 1967, this technique was applied to the SM–1A, using the nuclear core as the heat source. The anneal was performed at 572°F for 144 hours and 48 percent recovery was achieved. The problem itself and the method of recovering reactor vessel strength proved an important advance for the whole nuclear industry. Combined with the replacement of the SM–1A's steam generator in 1969, the plant was able to remain in operation until 1973.[26]

At the same time that the Army's Nuclear Power Program showed signs of shrinkage in number of plants on-line, it was also undergoing a reorganization. The Chief of Engineers conducted a Command Inspection of the Army Engineer Reactors Group on January 15 and 16, 1968, to study the whole nuclear program, as well as to determine the possibility of combining administrative functions and other areas of activity. The resulting investigation determined that the Nuclear Power Field Office Headquarters should be eliminated. Its functions would be performed by the Army Engineer Reactor Group Headquarters itself. The Report also suggested that fiscal, supply, and personnel functions be consolidated in a centrally located Administrative Services Department.[27]

In response to the inspection, on July 21, 1968, the Engineer Office approved the reorganization. The order established an Operations Division for the purpose of combining similar functions under one head. This new division included as subelements the organizations then designated as the Idaho Nuclear Power Field Office, the SM–1A Branch, Interservice Detachments, and the 535th Engineer Detachment.[28]

Reorganization of the Nuclear Program had little to do with its continuing activity. On July 16, the Chief of Engineers, Lt. Gen. William Cassidy wrote to the Army's Chief of Research and Development, Lt. Gen. Austin W. Betts, to discuss the future of the entire Program. Lt. Gen. Cassidy began by noting that budgetary limitations and other Research and Development requirements had resulted in a steady reduction of support for the Nuclear Power Program. Since 1963, funding had decreased from $7,580,000 to the present $2 million for FY 1969. Furthermore, the possibility still existed that an additional 20 percent reduction in funding might occur. In fact, the Chief of Engineers explained that a funding level of $2 million was the minimum at which a viable research and development program in nuclear power could be maintained.[29]

Gen. Cassidy reported that the original FY-1969 budget allocation had been only $1,400,000, which meant a cut of $1,150,000 over FY 1968. At that level, the Chief of Engineers said that his Office had been forced to seriously "consider recommending cancellation of the program or reducing it to a small study group effort." Ultimately $600,000 of the reduction was restored, which allowed for a minimal research and development effort to continue. However, Gen. Cassidy observed that if the program lost the 20 percent funding that was still under consideration, it would be left "in a precarious position." Consequently, the Chief of Engineers recommended "that the military need for nuclear power be carefully and thoroughly reevaluated." Specifically, he suggested to Gen. Betts that the Army Scientific Advisory Panel "establish an ad hoc group to study this program and... make appropriate recommendations as to its continuing nature and scope, or its abandonment."[30]

Gen. Cassidy recommended that the Study Group give consideration to several questions:

1. Were there valid potential applications and requirements for nuclear power plants (fixed, mobile, barge-mounted, hardened, or portable) in the next 10–15 years;
2. What technological, safety, economic, logistic or other factors will limit or prohibit the use of nuclear power plants for military application;
3. What level of research and development (manpower, funds, facilities, etc.) will be required to overcome the barriers to effective military use of nuclear power in the next 10 to 15 years;
4. Should the DOD organization structure of research and development in nuclear power (i.e., Army Nuclear Power Program, U.S. Army Engineer Reactors Group, and Army Reactors, U.S. Atomic Energy Commission) be changed?[31]

Gen. Cassidy noted that there had been numerous studies of the Nuclear Power Program in recent years, but said that none had addressed the overall problem as he now proposed. Nevertheless, he thought the earlier reviews would provide valuable input to the Advisory Panel Study. The Chief of Engineers concluded, "On the basis of the results of this study the Army must decide whether it wants or can afford a worthwhile nuclear power program. If the answer is negative, we should eliminate the program and apply the effort to other potentially more productive areas. If the answer is affirmative, the program must be funded at a productive level."[32]

In response, on August 3, Gen. Betts agreed with Gen. Cassidy that the Army needed to "take a hard and thorough look" at the program and so had initiated action to establish an Ad Hoc Study Group under the Army Scientific Advisory Panel. He also informed the Chief of Engineers that the

Nuclear Program had in fact had its funding reduced to $1.4 million for FY 1969. In light of Gen. Cassidy's belief that the Program could not successfully operate at that level, Gen. Betts indicated that the Army should obtain the results of the study as soon as possible and set a two-month period for conducting the analysis.[33]

Despite Gen. Betts' hope that the Ad Hoc Group would be functioning by early September, Dr. Theodore Taylor did not accept the invitation of the Assistant Secretary of the Army for Research and Development to be chairman of the Ad Hoc Group until November 14. Taylor was then President of the International Research and Technology Corporation in Washington. Subsequently, Dr. Carlyle J. Roberts, Director of the School of Nuclear Engineering at Georgia Institute of Technology; Dr. James J. Renier, General Manager of Advanced Communications Systems at Honeywell, Incorporated; and Dr. C. Rogers McCullough, Technical Director of Southern Nuclear Engineering all accepted invitations to serve on the group. The first meeting was held on February 10, 1969.[34]

At the meeting, Gen. Betts welcomed the Group and set forth the objectives for the study. The Office of the Chief of Engineers then provided a long briefing that included a history of the Army's Nuclear Power Program, its current status, and the current research and development program. Particular emphasis was placed on the Army's overall energy needs in the field. Although electrical power represented only a fraction of this requirement, the briefing officer pointed out that this was still "a staggering amount of energy. The demand for electrical power has grown rapidly during the past few decades and it is forecasted to continue to grow in the future." Nevertheless, the Army faced severe problems in supplying this electricity "due to the multiplicity of makes and models of existing electrical power generation equipment." At the same time, the commentator observed that the Army considered only its tactical needs in the field "to the virtual exclusion of a consideration of the needs of the forces that back up the tactical units." According to the officer, the Nuclear Power Program had given "a measure of the power demand for these backup forces."[35]

In defining the place of the program within the military's "technologically under-developed energy area," the briefer said the people in the Nuclear Power Program felt they had made a great contribution to improving the situation. Nevertheless, he candidly observed that the Army's Nuclear Power Program has "suffered continually from a lack of firm requirements. Essentially, as has been said before, nuclear power is a solution in search of a problem. Certainly this is an apt description of the ANPP. One of the major program efforts has been to try to define the need for nuclear power. Intuitively, we in the program feel that there is a need, but we have been unsuccessful to date in conveying it to the rest of the Army—or the the other services—or DOD."[36]

After suggesting that a large part of the problem has been the Army's inability to define its own requirements, the briefer closed by indicating four areas warranting sustained effort if the Army decided to continue with the Program:

1. Continuation of the research and development program in support of the existing plants, to find and correct problems, and thereby learn how to avoid similar problems in future plants;
2. Continued and increased efforts in the area of barge-mounted plants to make them better, more reliable, and more cost-effective;
3. Continued research and development pertaining to underground installations against the day when a firm requirement for power for these sites was identified; and
4. Continued efforts in advanced technology both of reactors and power conversion systems that offered the potential of significant advances in capability of the nuclear plants, which may make them attractive for general Army use, and which may allow nuclear power to help fill the demand for energy by the Army in the field.[37]

According to the briefer, the funding for such a program should be enough to maintain a viable research and development effort over a period of years without excessive cost. A funding level of around $4 million was considered adequate to maintain an active program with some hope of progress. At the same time, the briefing officer advised that there would have to be some provision for occasional increases in funding to allow for hardware procurement. Nevertheless, because of the spareness in recent research and development funding, the Program recognized it "could not jump in FY 1970 to a $4 million dollar funding level and still spend the money profitably." Consequently, the Program was asking for approximately $1.5 million for FY 1970 to "allow time for planning a coherent program while continuing current efforts," with the $4 million level reached in FY 1972. According to the briefer, this "programmed increase would allow adequate time for planning to assure maximum returns on the investment."[38]

The balance of the briefing sessions covered new technology, discussions with the AEC's Division on Reactor Development, and a search "to uncover some well defined military requirements for nuclear power." Between February and the beginning of April, members of the Ad Hoc Group individually reviewed information provided to them by the Group's military staff assistant. Its second series of meetings took place in Washington on April 2 and 3, and included discussions on the cost of conventional fuels, Minuteman Power Sources, the Silo Program, and a discussion of conclusions based on the investigation up to that point. A draft of the Final Report was prepared between April 2 and May 26, and circulated to members of the Group for comment and signature between May 28 and June 13.[39]

The Final Report concluded that current and projected nuclear reactor systems of the type that will meet military requirements could not be made cost competitive with conventional power sources except in selected remote locations. The Ad Hoc Group found that no firm military requirements for additional nuclear power sources existed despite the advantages offered by these systems. Nevertheless, the Group did conclude that thorough and creative examination of future military operations and weapons systems concepts was likely to uncover valid potential military applications and requirements for nuclear power sources. Therefore, the Final Report recommended that the Nuclear Power Program research and development effort eliminate experimental and hardware development efforts. It suggested that the Program's exploratory development should be reoriented toward a study group effort. Finally, it advised that the Army should undertake a study to evaluate the use of nuclear power sources in the context of future land force operations.[40]

The Army's Office of Research and Development asked the Corps of Engineers to review the Final Report and respond to a series of questions no later than August 12, 1969. These included: (a) the impact of the Study Group recommendations on current Nuclear Power Program operations, recommendations on the mission, function, concept of operations, and the optimum organization and annual funding for a nuclear power study group such as the one suggested in the Final Report; (b) recommendations on the objective and scope of the studies that could be undertaken by the nuclear power study group; and (c) submission of an implementing plan to phase in the study group effort and to phase out the current experimental and hardware development program. Because it seemed possible that the entire research and development program might be terminated in FY 1970, the Army also asked the OCE to comment on the impact of termination of nuclear power research and development funding on the Army Nuclear Power Program operations and to indicate funding required and a detailed plan to phase out the nuclear power exploratory development program.[41]

On August 15, Lt. Gen. Frederick J. Clarke, the new Chief of Engineers, responded to the Final Report. He concluded that implementing the recommendations of the Ad Hoc Group would effectively terminate the Army's research and development efforts in the nuclear power field. It would also require a reduction in force and/or transfer to other mission-related functions of 42 out of the 54 personnel now carried on the research and development payroll. There would be little effect on existing contracts because earlier budget cuts had already reduced new work "to an almost non-existent level." The recommendations would have no effect on the Army's in-house experimental program because it was in the final process of termination without regard to the Study Group recommendations.[42]

The Chief of Engineers rejected the recommendation that a study group be created within the Army Nuclear Power Program. He explained that studies such as that recommended "are best accomplished by the Combat

Developments Command (CDC), with technical support from OCE, as appropriate." In regard to the current operating power plants, Gen. Clarke recommended that "in view of the large investment already made in the ANPP," the existing plants should remain on-line until such time as major problems force them to be shut down. In this light, Gen. Clarke indicated that work within the Nuclear Power Program other than for plant operation, training, and support activities should be limited to:

1. Review and evaluation of future Army tactics, strategy, and doctrine as developed by the Combat Developments Command to determine if nuclear energy can provide a capability not otherwise attainable;
2. Operations research, systems analysis, and cost-effectiveness studies on deep underground and hardened power plants in order to determine potential applications, perform conceptual designs, and identify problem areas where research and development programs will be required;
3. Providing the interface between the Corps of Engineers and other DOD agencies on matters involving the application of nuclear energy technology; and
4. Maintaining close contacts with the AEC to assure that technology improvements developed by the AEC or by industry are included in the Army's efforts.[43]

Following a review of the Ad Hoc Study Group Report, on March 19, 1970, the Director of Defense Research and Engineering advised the Assistant Secretary of the Army that the Defense Department had approved a study program for the Army Nuclear Power Program not to exceed $500,000 a year. The Memorandum set out a general guide to the Nuclear Power Program stating that no hardware development programs should be initiated without prior program approval. The study program should concentrate on a search for potential future applications where nuclear energy could supply a needed capability otherwise unattainable. It should maintain a knowledge of promising nuclear reactor technology through liaison with the Atomic Energy Commission and the appropriate elements of the reactor industry. It should also seek application of cost-effective techniques to promising concepts when a substantial increase in needed capability is evident. The Memorandum also indicated that this study program would be reviewed periodically to determine future courses of action.[44]

In turn, the Army Research and Development Office informed the Chief of Engineers of the decision on April 3. The Engineer Office was requested to "reorient the nuclear power R&D program. In-house and contractual experimental work, such as the [Pulse Reactor Core Improvement Project] should be terminated. No hardware development program should be initiated without prior approval of this office." The Office also provided the Engineers with the guidelines from the Defense Research and Engineering Office.[45]

On June 11, the Chief of Engineers advised the Army's Chief of Research and Development that the Nuclear Power Program had not initiated any new experimental work or hardware development programs since the Ad Hoc Group had submitted its Report the previous June. With one exception, he said that all ongoing programs had been terminated or would be terminated when current contracts ended in September. The nil-ductility transition temperature testing and surveillance program, which supported the SM–1A pressure vessel annealing and the planned annealing of the SM–1A pressure vessel was expected to be completed in FY 1973.[46]

The OCE outlined the studies planned for FY 1971 and FY 1972, which the Combat Development Command would support with a 10-man study group with $500,000 funding in for the first year and $300,000 for the second. Other DOD agencies were expected to supplement the effort with men and funding to perform requested studies. The core group would be used to maintain an in-house state-of-the-art technical capability and to follow government and industry development in reactor concepts, energy conversion, and heat rejectors. It would review and identify steam heating, as well as mechanical and electric power requirements for the Army in areas where the use of nuclear energy should be beneficial. It would review future Army tactics, strategy, and doctrine, which the Combat Development Command would formulate in order to determine whether nuclear energy could provide specific capabilities not otherwise attainable. It would perform conceptual designs for nuclear power plants that had potential applications in the near future. It would also develop a cost-effectiveness analysis capability and the criteria for such analysis based on DOD and CDC guidelines. The OCE explained that in performing these functions, the study group "will identify potential applications, perform conceptual design and evaluate these concepts based on cost-effectiveness analyses. Contractual support to supplement in-house capability will be held to a minimum. No hardware development and experimental testing programs will be initiated without prior approval from OCRD."[47]

When the Nuclear Power Program lost its research and development capability, it became an organization without a mission, living on its past achievements. Even though four power plants remained in operation, the Program's personnel faced an uncertain future after years as part of an elite organization. Responding to the need for "a clear definition of the roles" of its Program's current components, the Program's Director, Lt. Col. Harvey Arnold, Jr., sent a Memorandum to all personnel on September 21, "in order to alleviate some misunderstandings which have developed."[48]

Acknowledging that the Program's mission was "not as clearly assigned as we would like it to be," Arnold indicated that the Corps of Engineers was formulating a new mission statement. In the meantime, he analyzed the current situation and anticipated the Program's future direction. On the negative side, Arnold summarized the limitations and prohibitions on the Program's research and development activities. He explained that the Engineer Reactor Group (ERG) would continue to train personnel to meet its requirements, provide technical support to its own plants, and to other agencies as requested, operate its own safety programs, and maintain the remaining nuclear plants and the powerbarge *Weber*.

According to Arnold, the loss of the research and development activity was countered by the additional mission of operating and maintaining the powerbarge *Weber* in the Canal Zone. Heralding the change in the organization's direction, Arnold said he expected the Corps of Engineers to give the ERG the new responsibility of operating other conventional powerbarges and diesel-electric generators. He explained that the Chief of Engineers now owned the oil-fired powerbarges *Inductance* and *Impedance*, which were then being operated in Okinawa by contract crews. The Director expected the agency to be given the mission of managing their deployment under normal conditions and being prepared to manage them for contingency missions.[49]

In regard to the operation of other conventional generating equipment, Arnold explained that the Chief of Engineers was considering a proposal to establish and maintain a contingency reserve of nontactical diesel-electric generators. In light of the Program's experience with electrical power-generating equipment, Arnold expected that it would participate in establishing this stockpile, acquired from excess generators in Vietnam and other areas. The Program would manage the maintenance of the machines and coordinate with and give technical assistance to the users.[50]

In fact, the Corps of Engineers took almost exactly a full year to formalize the new mission of the Nuclear Power Program. On October 13, 1970, the Chief of Staff of the Army assigned the Engineers the program management responsibility for the establishment and maintenance of an insurance level of nontactical generators. Following this up, on November 27, the Army Deputy Chief of Staff for Logistics, at the request of the Chief of Engineers, notified Army commands that the Engineers required 500-kilowatt and 1,500-kilowatt generators and associated distribution equipment that had been declared excess, in order to provide a stockpile of material for its new mission.[51]

In recognition of this "changing mission assignment," in a meeting on January 6, 1971, the Chief of Engineers concluded that the Army Engineer Reactor Group should become redesignated the U.S. Army Engineer Power Group (EPG). As a result of the meeting, the Power Division, the new designation for the Nuclear Power Division, was made "the central contact and coordination point within military construction for all power systems regardless of their type or application." And, by the beginning of March, work was progressing on the drafting of an Army regulation to define procedures and interagency agreements for the new Engineer activity.[52]

Meanwhile arrangements had been completed between the Chief of Engineers Office and the Army's Military Construction Agency for it to administer the retrograde, storage, and rehabilitation of excess Vietnam-used generators for the Engineers. Tooele Army Depot in Utah was designated as the site for the rehabilitation work and the storage of the equipment. With this assignment in hand, the Directorate of Military Construction published and circulated the inspection criteria to be used in examining equipment being considered in retrograde. In its notification, the agency noted that the Army's best long-term interest dictated that only economically reparable equipment should be turned over for rehabilitation.[53]

In pursuit of the Engineer Power Group's new mission, the Chief of its Operations Division visited Vietnam in order to make an overall survey determination of the nontactical generator systems and equipment suitable for retrograde. Concurrently, detailed planning for the establishment of equipment and associated component systems rehabilitation, system installation procedures, property accountability, parts inventory, and funding were under development. In addition, detailed planning was begun for formulating and writing a contingency employment plan for the *Inductance* and the *Impedance*. By March, however, the Engineer Power Group had received no official notification of availability of material although it did have information about potential surplus equipment.[54]

In any event, on September 22, 1971, the Chief of Engineers finally changed the name of the Engineer Reactor Group to the Engineer Power Group, effective October 1. The name change notwithstanding, the transition from an elite, research and development organization to essentially a reactive, emergency support agency required several years. Both internal structure and the handling of outside requests for loan of equipment had to be worked out by the Chief of Engineers and his Command Staff. In particular, Col. Arnold, now Chief of the Power Division, formulated an 18-point set of guidelines on the procedures and responsibilities for the operation of the Nontactical Generator Program. Issued on December 23, 1971, the Memorandum dealt with such matters as handling requests, deciding whether the program could supply the needed generators, payment for use, and transportation of the equipment.[55]

As the Nuclear Power Program had developed to meet a requirement to supply electricity in remote areas, the Nontactical Generator Program had evolved to meet the need for greatly increased amounts of electric power for combat support equipment. In Korea and to a far greater extent in Vietnam, new electronic technology required readily available, medium-voltage electrical-generating and distribution equipment and the trained operating teams to respond promptly to military demands. Although through its research and development efforts the Nuclear Power Program did not create a small nuclear power plant capable of meeting this need, its efforts as an electric-generating organization had produced a cadre of trained personnel; this made it the logical choice to carry out the mission of procuring and maintaining a pool of small conventional power-generating equipment for military and civilian energy use. Its personnel had the expertise to deploy and install the electrical generators and operate them until local technicians, trained by the men in the field, could take control of the equipment.

By April 1972, the Nontactical Generator (NTG) Program had received three 1,500-kilowatt and sixteen 500-kilowatt generators and had sent them to the Tooele Army Depot to await rehabilitation. Estimates from the Army in Vietnam indicated that the NTG Program would have twenty-eight 1,500-kilowatt and forty-five 500-kilowatt generators in its possession by July 1973. The Program estimated that the first generators would be available for deployment by January 1973. In addition to the generators being retrograded in Utah, the NTG Program also had twelve additional 500-kilowatt plants being rebuilt at the Sharpe Army Depot, which were scheduled for deployment by the middle of 1973.[56]

As the NTG Program developed, however, the nuclear component continued to decline. The PM–3A at McMurdo Sound was decommissioned in 1972 after 11 years of service. In making its decision, the Navy announced that the reactor had run its useful course. Nevertheless, in its last years of operation, the plant had developed several malfunctions, including leaks in its primary system. Among the problems had been cracks in the containment vessel that had to be welded. As a result, there was concern that a radioactive spill had taken place and leakage had run into the crushed gravel used as the facility's protective shield.[57]

After the decommissioning of the PM–3A began, studies showed that some radioactive coolant had in fact been lost before the completion of the welding operation. At the same time, another report revealed that some radioactive effluents that had been diluted with brine to keep them below the maximum concentration permitted by the government had nevertheless become concentrated in the antarctic soil. Consequently, while Navy decommissioning procedures called for the removal of all contaminated structures, the Nuclear Regulatory Commission ultimately required that all the radioactive soil also be removed. As a result, nearly 14,000 tons of low-level irradiated soil were dug up from the PM–3A site and transported for disposal to Port Hueneme, a naval base north of Los Angeles; the last load did not arrive until early 1979. In the meantime, the dismantled reactor and spent fuel rods were shipped to a radioactive storage site in Georgia for burial.[58]

During the same time period as the PM–3A was being shut down and dismantled, both the SM–1A at Fort Greeley in Alaska and the SM–1 at Fort Belvoir were decommissioned. Despite the SM–1A's new lease on life after the annealing process and the replacement of the steam generator, the repairs had only delayed the end of the plant's useful life. In late 1970, the Chief of Engineers decided that the SM–1A had successfully completed its mission of demonstrating the feasibility of building and operating a nuclear power plant in a remote arctic environment. As a result, the Army Engineer Reactor Group was directed to begin planning its decommission no later than the end-of-life of the SM–1A's fourth fuel core in 1973. The actual termination of plant operation was to be established in light of the details of plant decommissioning, seasonal conditions in Alaska, the energy requirements of Fort Greeley, and the funds allocated for the work.[59]

As the first step in the decommissioning process, the Reactor Group completed an Environmental Impact Statement on August 17, 1971. The Impact Statement called for Fort Greeley itself to assume operational and maintenance responsibility for the nonnuclear portions of the SM–1A facility and to operate the secondary power system with steam from the post's adjacent diesel power plant. The decommissioning plan for the reactor called for an in-place entombment. The decommissioning crew would remove and ship to an off-site disposal locale the nuclear fuel and other related nuclear material, the control rods, the control rod baskets, parts of the control rod drive mechanism, material surveillance equipment, and other primary system equipment.[60]

The remaining components and materials in the vapor container were to then be decontaminated to the maximum practicable extent, somewhat rearranged, and entombed in concrete in the lower portion of the vapor container. The concrete was to serve the dual purpose of providing additional biological shielding for any induced radiation and of making the materials inaccessible without deliberate and extreme effort. The remaining open, upper portion of the vapor container would be radiologically safe. Nevertheless, the container would be sealed by welding the access door shut.[61]

The Chief of Engineers approved the Environmental Appraisal Report on January 4, 1972 and the Engineer Power Group completed the detailed decommissioning plan in March. Although the Army did not formally approve the plan until May 5, decommissioning began on March 14, 1972, when the plant was shut down for the last time. By the end of the year, decommissioning was approximately 75 percent completed.[62]

In the meantime, on June 2, 1972, the Chief of Engineers announced that the SM–1 would be deactivated prior to May 1973, after 16 years of operation. Unlike the several field plants whose primary missions were to produce electricity for military installations, the major role of the SM–1 had been to train operating crews for deployment to the reactor sites. With the shutdown of all facilities except the *Sturgis*, the continued operation of the SM–1 to train the limited number of personnel for assignment to the nuclear powerbarge could no longer be justified, particularly in the face of the rising costs of the original plant's operation and maintenance. Moreover, the Fort Belvoir facility had never been cost-effective in producing electricity for the base in comparison to the price at which the Army could then obtain power from VEPCO. Finally, if the Engineers had continued to keep the

SM–1 in operation, the Engineer Power Group would soon have faced the expense of modifying the plant to meet projected AEC safety standards.[63]

As it had done with the SM–1A, the Engineer Power Group initiated an Environmental Assessment of the decommissioning process of the SM–1. On December 1, 1972, the EPG proposed the same basic plans for decommissioning the Fort Belvoir facility that it was following with the SM–1A. Like the Fort Greeley plant that was to remain in operation, albeit as a conventional power plant, the SM–1 building was to continue to be used as a training facility, an office area, and a museum. Consequently, the report provided for a careful decontamination of the building and the sealing off of all radioactive areas.[64]

Following approval of the Environmental Assessment and subsequent decommission plan, the SM–1 was shut down for the last time on March 16, 1973, with four of the original crew members participating in the ceremony. The plant's training functions were taken over by the reactor simulator and fuel-handling training facilities at Fort Belvoir. Following the completion of the decommissioning process in April 1975, the Engineer Power Group dedicated the plant's Visitor's Center and began using the building as a museum, providing movies and exhibits to tell the history of the Army's first nuclear power plant.[65]

With the shutdown of the PM–3A, the SM–1A, and the SM–1, only the *Sturgis* remained in operation, although not strictly in a military capacity. With its nuclear mission down to a bare minimum, the Engineer Power Group found itself with only one assignment, the Nontactical Generator Program with the limited objective of rehabilitating and storing small, conventional power plants. At this point, events and circumstances came together to give the EPG new life. Its new assignment allowed it to use its expertise in the power field not only to maintain the NTG Program, but to broaden its activities in related areas.

## NOTES

1. Interview with Gen. Kenneth Nichols, April 24, 1963.
2. *Project Charter, Floating Nuclear Power Plants,* Oct. 15, 1963
3. Interview with Ernest Fortino, Nov. 20, 1980.
4. Interview with Col. James Lambert, Feb. 16, 1963; Nichols interview; *News Bulletin #1,* Sept. 7, 1961; *News Bulletin #2,* Nov. 17, 1961.
5. *News Bulletin #1,* Sept. 7, 1961; *News Bulletin #2,* Nov. 17, 1961. Additional information on the building of the MH–1 came from the *Stugis* files, Box 5, ANPP.
6. *News Bulletin #2,* Nov. 17, 1961.
7. *News Bulletin #5,* Feb. 15, 1962; *News Bulletin #10,* Aug. 7, 1962; *News Bulletin #12,* Nov. 15, 1962.
8. *News Bulletin #14,* Feb. 5, 1963.
9. *News Bulletin #15,* April 25, 1963; *News Bulletin #16,* June 21, 1963.

10. *News Bulletin #26,* Aug. 14, 1964.
11. *News Bulletin #27,* Sept. 15, 1964; *News Bulletin #28,* Oct. 15, 1964; *News Bulletin #29,* Nov. 15, 1964.
12. *News Bulletin #29,* Nov. 15, 1964; *News Bulletin #30,* Dec. 15, 1964.
13. *News Bulletin #30,* Dec. 15, 1964.
14. Ibid.; *News Bulletin #25,* July 31, 1964.
15. *News Bulletin #25,* July 31, 1964.
16. *News Bulletin #2,* Feb. 15, 1965; *News Bulletin #6,* June 21, 1965; *News Bulletin #7,* July 20, 1965.
17. *News Bulletin #9,* Dec. 27, 1965.
18. *News Bulletin #1,* May 22, 1966; Interview with Col. Robert Bunker, Dec. 20, 1979.
19. *News Bulletin #1,* May 22, 1966; *News Bulletin #2,* Sept. 1966.
20. *News Bulletin #3,* Dec. 15, 1966; Nuclear Power Field Office (NPFO), *Historical Activities,* Jan. 1, 1967–July 1, 1967.
21. NPFO, *Historical Activities. Baltimore Evening Sun,* July 25, 1967.
22. Office of the Chief of Engineers, Technical Liaison Office, April 26, 1968; Office of the Chief of Engineers, *General Orders,* no. 5, March 15, 1968, MH–1A, Box 5.
23. *The Panama Star and Herald,* July 27, 1968, MH–1A, Box 5.
24. Ibid.
25. U.S. Army Engineer Reactor Group, *An Introduction to the Army Nuclear Power Program,* March 1970, Box 5, FESA General File. Additional material on the *Weber* came from the "Weber" file, Box 5, ANPP.
26. PM–1 files, Box 6, ANPP; SM–1A files, Box 6, ANPP.
27. Office of the Chief of Engineers, Report to Director, Engineer Reactors Group, June 21, 1968; Director, U.S. Army Engineers Reactors Group, Response to Chief of Engineers, April 30, 1968, Box 3.
28. Office of Chief of Engineers, to Director, Engineer Reactors Group, June 21, 1968, Box 3.
29. Chief of Engineers Lt. Gen. William Cassidy, Letter to Army Chief of Research and Development Lt. Gen. Austin Betts, July 16, 1968, Box 3.
30. Ibid.
31. Ibid.
32. Ibid.
33. Chief of Research and Development Lt. Gen. Austin W. Betts, to Chief of Engineers Lt. Gen. William Cassidy, Aug. 3, 1968, Box 3.
34. Ad Hoc Group on the Army Nuclear Power Program, *Final Report,* June 1969, Box 4.
35. "Briefing on Army Nuclear Power Program Past–Present–Future" (Presented to an Ad Hoc Group of the Army Scientific Advisory Panel, Feb. 10–11, 1969), Box 4.
36. Ibid.
37. Ibid.
38. Ibid.
39. Ad Hoc Group, Box 4. *Final Report.*
40. Ibid.; Army Office of Research and Development, to Chief of Engineers, July 18, 1969.

41. Army Office of Research and Development to Chief of Engineers, July 18, 1969.
42. Chief of Engineers Lt. Gen. F. J. Clarke, to Army Chief of Research and Development, Aug. 15, 1969.
43. Ibid.
44. Director of Defense Research and Engineering, to the Assistant Secretary of the Army (R&D), March 19, 1970.
45. Army Office of Research and Development, to the Chief of Engineers, April 3, 1970.
46. Office of Chief of Engineers, to Army Chief of Research and Development, June 11, 1970.
47. Ibid.
48. Lt. Col. Harvey Arnold, Jr., Memorandum for all ANPP personnel, Sept. 21, 1970.
49. Ibid.
50. Ibid.
51. Engineer Power Group, *Historical Activities Report, FY 71,* Feb. 10, 1972, hereafter cited as *Activities Report, FY 71; Status Review, Major Goals of the Chief of Engineers,* March 2, 1971, Box 4 (hereafter cited as *Major Goals*).
52. OCE, *Memorandum for Record,* Jan. 11, 1971; *Major Goals.*
53. *Major Goals.*
54. *Activities Report, FY 71,* Box 3.
55. Col. Harvey Arnold, *Nontactical Electrical Power Generator (NTG) Program,* Dec. 23, 1971.
56. Notes from the Deputy's Desk, April 4, 1972, Box 6.
57. *Washington Post,* Dec. 31, 1978, p. A2.
58. Ibid.
59. *Environmental Impact Statement for the Decommissioning of the SM–1A Nuclear Power Plant, Fort Greeley, Alaska,* Aug. 17, 1971, Box 6.
60. Ibid.
61. Ibid.
62. Army Engineer Power Group, *Historical Data—1972.*
63. Office of the Chief of Engineers News Release, June 2, 1972; *Environmental Assessment Decommissioning and Conversion of the SM–1 Nuclear Power Plant,* Dec. 1, 1972, Box 6; hereafter cited as *SM–1 Assessment.*
64. *SM–1 Assessment.*
65. Ibid.; *The Castle* (in-house newsletter of the Corps of Engineers), March 21, 1973, p. 8.

CHAPTER 8

# FACILITIES ENGINEERING SUPPORT AGENCY (FESA)

As the Engineer Power Group had become an agency in search of a mission by the end of 1973, the Corps of Engineers was seeking an agency to undertake a new assignment. In addition to its civil and other military engineering assignments, since the end of World War II the Corps had supplied Engineer officers as post engineers to operate utilities and maintain facilities on Army bases. For the most part, in the decades of the 1940s, 1950s, and 1960s, the officers assigned to these positions were not of the caliber of those who had or were to later receive the important assignments necessary to advance to positions of high command within or outside the Corps.[1]

Given this reality, the assignment as a post engineer seemed to be a "kiss of death" to a career even if the records showed the officer's performance as post engineer rated as high or higher than his other assignments. In addition to any lack of talent, the post engineer seldom had an adequate staff in quality or numbers to provide him with the professional assistance he needed. When these shortcomings were combined with the lack of resources in funds and engineering support from the Corps itself, a low level of post maintenance usually resulted. Such deficiencies in men and actions gave rise to a bad image for the Corps within the Army itself.[2]

In 1968, in recognition of this and related problems, the Army Chief of Staff had initiated a study to provide a comprehensive analysis of the total management of the service's real property maintenance activities. Lt. Gen. L. G. Lincoln headed the study group that examined all reports of previous investigations and did its own research and analysis. Unlike the earlier studies whose recommendations "were rarely if ever carried out," according to Lincoln, his work gave new impetus to the operation and maintenance of the Army's facilities. A subsequent study described the Lincoln recommendations as giving "new direction and impetus" to the Army's operations and maintenance of its Real Property and "a ten year benchmark" for subsequent analysis.[3]

To deal with the need to provide engineering support for post facilities, the Lincoln Study Group recommended that the Army allocate funds to be controlled by the Chief of Engineers. This money would then be further allocated to districts, which would enable facility engineers to obtain engineering assistance without payment from their own resources. In regard to the facilities engineer, as Gen. Lincoln and his Study Group chose to rename the position of post engineer, the Study Report recommended that the selection of the officer be made on the same basis as any Engineer officer being assigned to a Corps of Engineers district. Not only would the man be "one of the best," according to Lincoln, but he would also subsequently be given the same consideration in personnel evaluation as an officer who had served in a district. The ultimate effect of this change would be to provide a better caliber of officer who would then be perceived in a better light by both peers and superiors, would have a reason to make a record for himself while in his assignment, and would be able to acquire engineering expertise that would be useful in his later career.[4]

To accomplish an assignment, however, the facilities engineer would also need support from an organization capable of providing broader assistance in emergency situations such as the breakdown of electrical-generating equipment and heating plants. In implementing the Lincoln Study Group proposals, Lt. Gen. W. C. Gribble, the Chief of Engineers, had the logical candidate to assume the assignment: the Engineer Power Group had capabilities he knew well after a long relationship.

Gribble had joined Col. Lampert and Maj. Bill Taylor in September 1953, during the research and start-up phase of the Army's Nuclear Program and had become the number-two man in the organization. He was later the Alaska District Engineer responsible for the construction of the SM–1A and, in the mid-1960s, had become the Director of the Army Reactors Group. Consequently, in response to Chief of Staff Gen. Creighton Abrams' order to implement the Lincoln Study and upgrade facilities on Army bases, Gen. Gribble knew where to turn. Moreover, he appreciated the possible resistance within the organization to the shift in mission to what many considered a less glamorous assignment. Nevertheless, with the Nontactical Generator Program its only significant mission, the Engineer Power Group was clearly in a situation where it had to take on additional activities, particularly if it was to maintain its individual identity in the Corps of Engineers.[5]

In announcing its new mission to the Engineer Power Group staff meeting on January 14, 1974, Maj. Mark Magnussen, then the Deputy Commander and Director of the Agency, explained that the organization would now be under the Facility Engineering Directorate in OCE instead of in the Military Construction Directorate. The change meant that the Power Group would "become more facility engineering oriented." To help set up the new assignment and its operation, Magnussen told the staff that the Chief of Engineers was sending to the headquarters Bill Taylor, Col. Lampert's assistant during the birth of the Nuclear Program, to study the current state of affairs as a consultant and make recommendations.[6]

Gribble, the Chief of Engineers, had initiated the changes in organization by deciding to create the Facilities Engineering Directorate within the Corps of Engineers headquarters organization, elevating it from division status with the Military Construction Directorate. Up to that time, Military Construction had the responsibility for both the Army's new construction and for the operation and maintenance of Army facilities. By making the Facilities Engineering Directorate a separate entity, Gribble thought it would be better able to cope with the complex requirements of facilities engineering on the Army's installations worldwide.[7]

The first director of the new Directorate, Brig. Gen. William Wray, had been Officer in Charge of the initial SM–1 crew that operated the plant under ALCO's supervision. He had taken part in the original nuclear engineering training program, had worked with Lampert and Gribble in establishing training procedures for power plant crews, and then served as the Nuclear Power Program's first Chief of Operations and Training. As a result, he was sensitive to the changes the Agency had undergone through the years. From Gribble, Wray received guidance about areas on which he should concentrate in his new assignment, particularly in applying the talent that existed in the Engineer Power Group to the problems faced by the facility engineers on Army installations. With Bill Taylor's help in examining the Agency and with his recommendations, Wray reorganized the Power Group and gave it a new mission.[8]

Organizationally, the execution of Facilities Engineering on Army bases did not come under the Chief of Engineers' direct control. However, the Chief did have Army staff responsibility for development of plans and programs for facilities engineering operations of the Army. These programs are referred to as the Army's Real Property Maintenance Activities. As a result, the Facilities Engineering Directorate executed the Chief of Engineers' Army Staff responsibility of providing staff supervision and overall evaluation of the facilities engineering functions. These included, but were not limited to, maintenance and repair of real property, urgent minor construction, operation of utilities plants and systems, fire prevention and protection, and all related engineering services.

The Directorate had several additional functions, such as design of the Integrated Facilities System—a multicommand, automated information and evaluation system encompassing the life-cycle management of real property resources from conception, through design, construction, operation, maintenance, and disposal. It also prepared quarterly reports on air and water pollution control, supervised the Army's Nontactical Power Program, executed the responsibility of the Chief of Engineers as the designated Army

Power Procurement Officer, and participated in the Secretary of the Army's energy program.

During the spring of 1974, the Engineer Power Group worked to adjust to its new organization and mission. In addition to operating and maintaining the *Sturgis*, the Group continued its research efforts on several projects. These projects included one to recycle laundry waste water and another to develop a sewage treatment process using gamma irradiation. On May 20, 1974, the Engineer Office officially changed the name of the Engineer Power Group to the Facilities Engineering Support Agency (FESA) and on July 5 formally established and defined the mission of the new organization.[9]

The change in name mirrored the new Agency's shift in focus from a power-generating–oriented activity to a facilities support organization. The continued operation of the *Sturgis* and related nuclear activities occupied perhaps 5 percent of FESA's work and that only until the final shutdown of the powerbarge in late 1976. Although the NTG Program retained its own identity within the Agency, its generators served as an emergency power supply for Army facilities and other government requirements in peacetime as a means for maintaining its wartime readiness for the nontactical military purposes for which it was established.

In its new orientation, FESA was to function as the basic Corps of Engineers field agency "to support, with technical assistance, the Facilities Engineering operations at the installation level." It was also to provide requested technical assistance to the Facilities Engineering Directorate in OCE. In addition, the Agency was to conduct assigned studies on Facilities Engineering matters on post, camp, and station equipment and practices.[10]

The Agency retained, "within prescribed limits," its original program of research and development that was "applicable to nuclear power plants and other utilities systems using radioactive materials." It also continued to investigate the feasibility of and to develop programs for application of nuclear energy systems to meet the operational requirements of the military services, exclusive of marine propulsion and air and space applications. It was to conduct investigative studies related to electric power-generating equipment with prime responsibility for all nuclear systems and attendant responsibility for fossil-fueled fixed power systems.[11]

In this area, FESA was to serve as the basic Army organizational framework for the Army Nuclear Power Program, which continued to function as an inter-agency activity within the Defense Department. The new agency also became the Pentagon's link with the new Energy Research and Development Administration which had taken over many of the AEC's functions when the Commission was disbanded in 1974. In this capacity, FESA was to operate and maintain the Corps of Engineers nuclear power plant and designated floating fossil-fuel plants. The Agency was to continue the NTG Program with its mission "of establishing and maintaining a contingency reserve of nontactical mobile generators and associated distribution systems." In this assignment, FESA was "to manage a peacetime loan program to maintain operational readiness of equipment and support capability."[12]

In addition to its responsibility for Army equipment, the Agency was to serve as the organization responsible for the readiness, operation, maintenance, and periodic testing of designated transportable power-generating equipment controlled by the Joint Chiefs of Staff. To do its work, FESA was to carry out a training program to support the remaining nuclear power plant as well as the NTG programs. In this capacity, it was to define and establish or coordinate with the Army and other user agencies, training programs to qualify operating and maintenance personnel for the users of floating and land-based NTG equipment.[13]

FESA was to participate in the Army's Reactors Health and Safety Programs and provide technical support and consulting services in nuclear-related and power plant technology to other services and federal agencies as requested. Finally, upon request, the Agency was to provide to other Army components engineering and operational support, technical advice, direct and general support maintenance and training relating to the installation, operation, and maintenance of NTG electrical power plants and distribution systems.[14]

Even though FESA's work was to focus on conventional power plants and facilities support, the Agency did continue to use its nuclear expertise, at least initially. On June 12, 1974, the Assistant Secretary of Defense (Installations and Logistics) sent a Memorandum to the Secretary of the Army directing the service to take the initiative in a tri-service re-evaluation of nuclear power plants for Defense Department use. This renewed interest at the very same time the last of the Army's reactors were being decommissioned reflected the sudden rise in fuel costs resulting from the 1973 Arab Oil Embargo. In addition, the military was responding to the general uncertainty in availability of fossil fuels and the projected overall increases in armed forces energy consumption. At the same time, with the energy crisis putting a heavy burden on commercial utilities to meet demands in the private sector, the military needed to assure itself of a reliable uninterrupted source of energy.[15]

In August 1974, FESA's re-evaluation plan called for the formation of an Ad Hoc Study Group to conduct a one-year evaluation of the feasibility of using nuclear power plants to meet DOD installation energy requirements. This plan received approval from the Defense Energy Policy Council. The study group was composed of representatives from each Service and was chaired by the Director of the Army Nuclear Power Program. During its work, the Study Group investigated several small reactor concepts that offered potential application as central power station plants for support of DOD facilities.[16]

The first phase of the study was directed toward identifying the economic and environmental constraints associated with the use of nuclear power.

The results of the Study Group indicated that nuclear power remained a viable alternative to continued dependence on commercial fossil fuel energy sources. Nevertheless, the Department of Defense took no steps to implement any of the plan's proposals.[17]

While FESA was studying possible future military uses of nuclear energy in 1974 and 1975, it continued its day-to-day involvement with nuclear power through its operation of the *Sturgis*. In 1973, the Engineer Power Group had informed the Panama Canal Company that no more nuclear fuel would be purchased for the powerbarge but the cores it had on hand would be sufficient to operate the *Sturgis* through 1976. At the same time, however, the Group informed the Canal Company that both the *Sturgis* and the *Weber* could be shut down sooner due to a major mechanical breakdown, a tightening of safety requirements, or severe manpower restrictions.[18]

On its part, in June 1974, the Canal Company indicated to FESA that it would like to retain use of the *Sturgis* through 1980. On September 2, 1974, however, the *Sturgis* was shut down for refueling and installation of an interim Emergency Core Cooling System. In February 1975, following zero power experiments and evaluation of two prototype control rods, the plant went back on-line. However, it was limited to 50 to 60 percent of its operating capacity because of the interim nature of the emergency core cooling system. Final installation of the system was scheduled for the summer of 1976, at which time the *Sturgis* was to resume full power operation.[19]

By late 1976, however, the United States was in the midst of negotiations with the Panamanian Government to turn the Canal over to Panama. It was decided that both the *Weber* and the *Sturgis* represented very inviting targets for radical elements in the population; the decision was made to remove both power plants from the Canal Zone and return them to Fort Belvoir before redeploying them. The *Sturgis* was already in the midst of a shutdown and had only two weeks to stow all equipment and prepare for its towing mode.

Off the North Carolina coast under tow on the evening of January 6–7, 1977, the *Sturgis* ran into a severe winter storm. Early in the morning, the Coast Guard directed evacuation of the seventeen-man tow crew by helicopter, which flew the men to Wilmington, North Carolina. That night a recovery team from FESA flew to Wilmington to organize and conduct reboarding operations. After an airborne radiation survey that showed no leakage, the FESA team was lowered by helicopter to the barge in midmorning on January 8 and made a damage assessment. The *Sturgis* was towed to the Military Ocean Terminal at Sunny Point, North Carolina for temporary repairs and then sent on to Fort Belvoir.[20]

By the time the *Sturgis* reached Fort Belvoir, FESA had received word that on January 5, the Chief of Engineers had informed the Army Chief of Staff of his intention to decommission the plant in the absence of a recognized operational requirement and the high cost of continued maintenance and operation. The OCE had made a comprehensive study to determine the economics of operating the powerbarge for another five years while also conducting a survey to determine where it might be re-sited. When no suitable locale could be found, the Chief of Engineers ordered the *Sturgis* decommissioned. The same process used with the land-based plants was put into action. Following the completion of the job in early 1978, the *Sturgis* was placed in anchorage in the James River Reserve Fleet.[21]

With the last of its nuclear power plants shut down, FESA's power mission, under its NTG Program, remained to continue the pool of electrical generators and distribution equipment for peacetime emergency and wartime contingency use. This work included the deployment, operation, and maintenance of numerous power plants in a wide range of locations both within and outside the continental United States—including Guam, Thailand, Samoa, Alaska, Okinawa, and Jamaica. FESA also became involved in the Corps of Engineers Research and Development Program, interfacing with the Energy Research and Development Administration. It received the assignment of the overall energy research and development responsibility for the Corps of Engineers. Its research and development efforts, in conjunction with its other engineering and technical skills, permitted the Agency to act as a technical transfer tube to the field in the area of energy and energy conservation.

In its support activity, FESA continued the heritage of the Nuclear Power Group's mission of supplying electric power in remote areas. In initiating the Army's Nuclear Power Program, Gen. Kenneth Nichols had seen the development of small, mobile, and portable nuclear reactors contribute to the advancement of nuclear power to generate electricity for peaceful purposes. In this light, it is ironic that one of FESA's major projects became supplying support equipment and power on Eniwetok from 1977 to 1980 to military personnel clearing up the radioactive debris and soil left by the atomic and hydrogen bomb tests that ran from 1947 through the late 1950s.

To carry out its mission, FESA provided the Joint Task Force's work force with a 2,000-kilowatt power plant and operating personnel. During the three years the generator supplied electricity for the clean-up, FESA had the opportunity to demonstrate the operation of the Army Functional Component System (AFCS) Program. This Program required the Army to prepare contingency designs for facilities it could use or might need in theaters of operation in various climates from arctic to tropic. FESA had received the assignment for the AFCS Program and then incorporated the NTG Program and its equipment into the AFCS Program.

To the extent possible, the 84th Engineer Battalion out of Hawaii followed the AFCS Program plans in constructing the power plant facility on Lojwa Island. This plant served the Army and the Nuclear Regulatory Commission as the command post from which to direct the clean-up operation throughout the test areas. The generator came from the NTG Program assets overhauled at Sharpe Army Depot in California and was delivered by the Navy to Lojwa. The plant, which went into operation in September 1977, enabled FESA to put the theory of the NTG Program into practice and the three-year mission provided significant input to its contingency planning aspects.[22]

As it entered the 1980s, FESA had received five specific missions from the Chief of Engineers. Overall, it served as a staff and technical support element to the Army. In this capacity, it assisted in the preparation of policy and guidance material related to the Nontactical Generator Program and Real Property Management System. Specifically, the Agency managed the Nontactical Generator Program with the mission of establishing and maintaining a contingency reserve of nontactical generators and associated distribution systems, including both fixed and floating power plants and high-voltage generation applications. Under this responsibility, FESA maintained a cadre of trained personnel to meet initial contingency requirements. In addition, it managed a peacetime loan program to maintain operational readiness of equipment and support capability.[23]

Furthermore, upon request the NTG Program provided to the Army, operational support, engineering and technical advice, direct and general support maintenance, and training relating to the configuration, installation, operation, and maintenance of electrical power plants and distribution systems. It identified, evaluated, and coordinated with the Department of the Army Readiness and Material Development Command (DARCOM) and provided design as requested for improvements in NTG Program equipment, methods of transportation, and methods of installation. It also provided to the Army Functional Component Systems Program Manager the operational requirements and reviews of its designs in areas of electrical generation and distribution. Pursuant to this mission, FESA implemented a training program to support the NTG Program.[24]

A second mission was to provide engineering surveys, consultant services, and other support to Army Major Commands (MACOMs) and installations, in technical and operational aspects of the Real Property Management System. The NTG Program also provided technical support, advice, consultation, and assistance to Army organizations; activities, commands, and installations in operation; maintenance, repair, and modification of real property facilities to include buildings, structures, family housing, grounds, roads, railroads, paved areas, water treatment plants, and distribution systems; waste water treatment plants and collection systems; heating and air conditioning plants; fuels; hospitals; electrical-generating plants and distribution systems; energy control systems; security systems; sewage systems; pest control; land management; fire prevention; and custodial services. Moreover, the Agency provided specialized technical assistance and centralized services such as infrared surveys of roofs and utility lines, boiler efficiency testing, boiler water treatment, electrical short circuit and coordination studies, load and power factor analyses, waste heat recovery studies, corrosion reduction surveys, and other programs as required.[25]

Related to this activity, FESA developed policy and established standards and procedures for utility contracting and procurement matters in the Army in addition to technical aspects of Army utility contracting practices and procedures. It monitored and maintained Army staff technical surveillance over utility contracting and procurement practices and procedures. A FESA staff member served as the Deputy Army Power Procurement Officer, developed procedures and standards, and administered the Army's Utility Contracting Program for the Chief of Engineers. In addition, the Agency provided DOD technical representation on utilities procurement matters with Government agencies and Congress.[26]

In the area of Systems Management, FESA served as the proponent agency representative for the design, development, testing, evaluation, training, and implementation of all Facilities Engineer–related management information systems. In the area of technology and energy, it provided analysis and planning for technological change to the Army and the Chief of Engineers. It conducted fixed facilities investigations in energy conservation, energy management, and the application of alternative energy sources. It also maintained the Energy Conservation Technical Information and Analysis Center and prepared the Army Facilities Energy Plan.[27]

FESA performed facilities investigations as well as test and evaluation of new technologies and products produced by laboratories and industry. These services included quick payback (less than a year) solutions in such areas as alternate energy source utilization, conversion systems, reduction of utility costs, and improvement of energy conservation. In this regard, FESA also provided technical monitorship for the Chief of Engineers of investigational programs in other areas and conducted special studies to establish guidance on technical facilities engineering matters.[28]

Finally, FESA managed the Army Reactors Health and Safety Program for the Corps of Engineers and had delegated proponency for dealing with the use of nuclear material. It also had responsibility for safety checks on the decommissioned SM–1 and the MH–1A, which included periodic environmental monitoring of the plants and taking any necessary corrective action to ensure nuclear integrity.[29]

Consequently, FESA stood as the direct connection with the original Nuclear Power Program, whose mission was to provide the Army with electrical power. To be sure, its research and development efforts had become limited and its mission less glamorous. Nevertheless, given the great need within the military for electricity and the maintenance of adequate facilities for Army personnel, the Agency perhaps served a more useful function—at least for the long haul.

## NOTES

1. Engineer History Office, interview with Lt. Gen. L. G. Lincoln, Jan. 15, 1980 and discussion with author, Dec. 7, 1981.

2. Ibid.

3. Ibid.; RPMA Study Group, OCE, Special Study Total Management of Real Property Maintenance Activities (RPMA), Dec. 1968, Box 7, ANPP; Department of the Army Study Group, "Real Property Management Activities (RPMA)," in *Executive Summary*, vol. 1, March 1978, Box 7, ANPP.

4. Lincoln interview; RPMA Study Group, Special Study Total Management of Real Property Maintenance Activities (RPMA), Dec. 1968.

5. Taylor interview, Dec. 21, 1981.

6. Minutes of Staff Meeting, Jan. 14, 1974; Taylor interview. All material on FESA is found in Box 5, ANPP records.

7. Gen. Wray interview, April 10, 1980.

8. Ibid.; Staff meeting, U.S. Army Engineer Power Group, Jan. 14, 1974.

9. *Power Group News*, Spring 1974; OCE, *General Order 19*, June 13, 1974, Box 5; U.S. Army Facilities Engineering Support Agency, Office of the Chief of Engineers, *Regulation No. 10–1–11*, July 5, 1974.

10. Office of the Chief of Engineers, *Regulation No. 10–1–11*, July 5, 1974.

11. Ibid.

12. Ibid.

13. Ibid.

14. Ibid.

15. *Tri-Service Re-evaluation of Nuclear Power Plants for DOD Installations*, July 11, 1974, Box 5; Lt. Col. M. H. Magnussen, "The Approaching Power Crisis," n.d., Box 2, ANPP; Lt. Col. M. H. Magnussen, "Supporting the Electric Army," n.d., Box, 2, ANPP.

16. Ibid.; *Project Summary*, Oct. 23, 1975, Box 5.

17. *Project Summary*, Oct. 23, 1975; Deskside Briefing for Dr. Walter LaBerge, Under Secretary of the Army on Army Nuclear Energy Application, July 26, 1979; Interview with Dr. Harold Hollis, Box 7.

18. *Quarterly Report on Implementation of ANPP Guidance*, April 9, 1973, Box 5.

19. FESA Staff Meeting, June 26, 1974, Box 5; FESA Notes, Feb. 18, 1975, Box 5.

20. FESA, *Annual Historical Report—1977*, Box 5.

21. Ibid.

22. FESA, *Annual Historical Report—1977*; Joint discussion between Col. Edgar Mixan and Gen. Lawrence Lincoln, Nov. 10, 1980 (untranscribed).

23. U.S. Army Corps of Engineers, *Regulation No. 10–1–11*, Aug. 14, 1981, Box 2.

24. Ibid.

25. Ibid.

26. Ibid.

27. Ibid.

28. Ibid.

29. Ibid.

# Bibliographical Essay

The Facilities Engineering Support Agency (FESA) made the research for this monograph relatively easy. As the Army's nuclear power program evolved through several incarnations, each organization saved its historical records and passed them along. Consequently, when FESA contracted me to write its history, it had in its immediate possession almost all the documentation needed to complete the project. Included in the material were a series of interviews from 1963 which Charles W. Lynch conducted for the History Office with the Engineer officers who had helped create the nuclear power program. The collection also included Kenneth J. Deacon's unpublished history of the nuclear power program up to 1972. I located a few additional documents at the Federal Records Center in Suitland, Maryland, and added copies to the FESA collection. In addition, FESA transcribed the interviews I conducted among former and current members of the organization and the transcripts became part of the collection. As a result, I had ready access to all relevant research material during the writing of the manuscript.

The Corps of Engineers Office of History assumed control of the research material once the manuscript was completed and in April 1986, under contract, the Office prepared a research guide to the collection. The general citation to all the material used in the manuscript reads: "The Army Nuclear Power Program, Research Collections, Office of History, U.S. Army Corps of Engineers, Fort Belvoir, Virginia." Specific note citations in this text are to the boxes in which the documents and interviews are located. The individual files have no reference numbers, however, and two or three files often have the same label, for example, "SM–1." Moreover, the documents within a file are often not in chronological order. As a result, a researcher may be forced to go through every document in each file to find the specific item he seeks.

The collection itself totals eight boxes and the two boxes of material which Dr. Deacon had assembled. The material is arranged "by subject, and thereunder approximately chronologically." In fact, copies of the same documents may be found under several subject headings in different boxes. Nevertheless, the guide does provide fairly reliable direction for a person doing research on the topic.

# INDEX

ALCO Products (ALCO), 26, 30, 32, 34, 50, 52–53, 59–63, 68–69, 73
Army Corps of Engineers, 2, 4–5, 7–8, 10–13, 15, 20–22, 24, 32–33, 36–38, 42–44, 50–51, 53–54, 57–59, 71–72, 79, 89, 95, 98, 106, 108–10, 112, 117, 119, 122–24
Army Engineer Power Group, 109–13, 117–20, 122
Arnold, Col. Harvey, Jr., 108–10
Atomic Energy Act of 1946, 5–7, 13–14, 16
Atomic Energy Commission (AEC), 3–16, 20–25, 29–30, 32–33, 37–38, 42–43, 45, 72–74, 76, 81–83, 85–88, 90, 92, 100–101, 107
Bacci, Maj. Joseph, 20, 30–31
Betts, Lt. Gen. Austin W., 102–4
Bunker, Capt. Robert, 99–100
Burns and Roe Company, 70
Camp Century, Greenland, 55, 57–64, 66, 68–71, 79
Cassidy, Lt. Gen. William, 102–4
Clarke, Lt. Gen. Frederick J., 106–8
Collins, Gen. J. Lawton, 4–7
Dean, Gordon, 9, 12
Dunning, Dr. John, 3, 5, 7, 9
Eisenhower, Pres. Dwight D., 11, 13, 15, 37
Facilities Engineering Support Agency (FESA), 117–25
Fort Greely, Alaska, 42–43, 47–54
Graves, Maj. Ernest, Jr., 35
Gribble, Col. William C., 21, 32, 42, 45, 50–52, 87, 118–19
Groves, Gen. Leslie, 1–2
Hafstad, Dr. Lawrence, 5, 7, 10, 13, 15, 21, 24
Hertford, Brig. Gen. Kenner F., 13–14, 25–26
Lampert, Col. James B., 8–16, 19–22, 24–26, 35–37, 45–46, 49–50, 81, 118–19
Lincoln, Lt. Gen. Lawrence G., 117–18
Magnussen, Maj. Mark, 118–19
The Martin Company (later Martin-Marietta), 58–59, 73–74, 76–77, 96–100
Military Compact Reactor (MCR), 87–89, 93
MH–1–A Nuclear Power Plant (Mobile High). *See* Sturgis
ML–1 (Mobile, low-power reactor), 89–93
Nichols, Maj. Gen. Kenneth D., 2–16, 22, 95, 123
Oak Ridge National Laboratory (ORNL), 13–16, 19, 21, 30–32
Office of Chief of Engineers. *See* Army Corps of Engineers
Pace, Frank, Jr., 4, 8–10, 14
Panama Canal Company, 100–101, 122

Peter Kiewit Sons' Construction Company, 50–52
Pick, Lt. Gen. Lewis, 2, 5–6, 8
PM–1 Nuclear Power Plant (Portable Medium), 55, 73–75, 77, 79, 102
PM–2A Nuclear Power Plant (Portable Medium), 55, 57, 61–74, 77–79
PM–3A Nuclear Power Plant (Portable Medium), 75–78, 111, 113
Quarles, Donald, 20, 22–24, 33
Rickover, Rear Adm. Hyman, 8, 10, 12, 21
SL–1 Nuclear Power Plant (Stationary Low), 79, 81–87, 93
SM–1 Nuclear Power Plant (Stationary Medium Prototype), 30–38, 42–43, 49–51, 81–82, 96, 99, 111–13, 125
SM–1A Nuclear Power Plant (Stationary Medium), 50–55, 57–58, 61, 75, 102, 108, 111–13
Stevens, Robert, 12, 14, 16, 22–23
Strauss, Louis, 11, 13–16, 23, 33, 37
Sturgis, Lt. Gen. Samuel D., 12, 49–50, 98
*Sturgis* (Nuclear Power Barge, MH–1A), 93, 95–102, 113, 120, 122–23, 125
Taylor, William B., 10–12, 20–21, 118–19
Truman, Pres. Harry S, 1–2
Virginia Electric Power Company (Vepco), 98, 112
*Weber* (conventional power barge), 101, 109, 122
Wilson, Charles, 15, 21
Wray, Brig. Gen. William, 119

**About the Author**

LAWRENCE H. SUID is a Washington-based military historian and independent writer. He is the author of *Guts & Glory*, *Scenes of Conflict*, and *The History of Armed Forces Radio and Television Service*.

www.ingramcontent.com/pod-product-compliance
Ingram Content Group UK Ltd.
Pitfield, Milton Keynes, MK11 3LW, UK
UKHW021906150926
475560UK00003B/218

9 780313 272264